Pocket Pal®

A Graphic Arts Production Handbook

Eighteenth Edition

INTERNATIONAL  PAPER

First Edition — 1934
Second Edition — September 1938
Third Edition — October 1954
Fourth Edition — November 1955
Fifth Edition — October 1957
Sixth Edition — April 1960
Seventh Edition — June 1963
Eighth Edition — November 1964
Ninth Edition — February 1966
Tenth Edition — May 1970
Eleventh Edition — December 1974
Twelfth Edition — March 1979
Thirteenth Edition — May 1983
Fourteenth Edition — May 1989
Fifteenth Edition — May 1992
Sixteenth Edition — July 1995
Seventeenth Edition — August 1997
Eighteenth Edition — August 2000

Copyright© 1934, 2000
by
INTERNATIONAL PAPER COMPANY
6400 Poplar Avenue
Memphis, TN 38197

foreword

Printing is undergoing the most important change in its history after being acclaimed the most important invention in the second millennium. Ever since its invention in 1450 A.D. printing has been the main medium of graphic communications. Now it must share that distinction with computerized digital technologies, mainly the Internet.

Since its debut in 1934, the *Pocket Pal* has been the authoritative introduction to the graphic arts for many artists, designers, publishers, advertisers, students and buyers of printing. Acclaimed by many as the best publication of its type, *Pocket Pal* is now in its 18th Edition as International Paper has continued to strive to keep its content current and accurate.

At the time of *Pocket Pal's* first edition, the term *printing* meant *letterpress,* and *lithography* and *gravure* were fledgling processes. Now the word *printing* encompasses not only all the graphic arts but also all the graphic communications, digital imaging and printing processes that output hard copy.

This 18th Edition is the sixth since 1983 when digital imaging beyond electronic typesetting and scanning began to replace prepress operations. Since the 15th Edition in 1992, updated in 1994, digital printing has been replacing conventional printing processes for short run printing and creating new means for on-demand and variable information printing. This edition attempts to put conventional and digital printing into proper perspective, and show what impact the new systems are having on the present and future uses and markets of printing.

The new information for this major revision has come from many sources and authorities. The editor thanks them for their contributions to this edition, in particular, Frank Romano, Chairman of the School of Printing, Rochester Institute of Technology, who revised the typographic section, updated the digital prepress section and helped with the new digital printing sections.

Michael H. Bruno
Graphic Arts Consultant
Editor

Contents

History

INTRODUCTION

Printing is a means of graphic communications. It is the reproduction of quantities of images mostly on paper which can be seen or perceived visually. Regardless of the great number and variety of printed products they all have one thing in common: each has a visible image produced in quantities of one or more.

Printing and publishing is a big business in the U.S. which altogether represents about 3 percent of the gross domestic product (GDP). Among all U.S. manufacturing industries, it ranks first in total number of establishments — more than 60,000 — with about 80 percent having fewer than 20 employees.

Today's printer owes much to the sciences, particularly electronics, computers, chemistry, optics and mechanics. Modern printing has become highly sophisticated. As new digital workflows with new prepress systems, lasers, plates, presses, inks, papers, electronic controls and digital imaging and printing systems have been developed, printing has been transformed almost completely from an art to a science.

THE EVOLUTION OF PRINTING

Man's earliest known attempt at a visual record of his life and times dates back 30,000 years. These were wall drawings called pictographs, superseded by the more complex ideographs. They, in turn, were succeeded by the Persians' cuneiforms, and then by hieroglyphics, perfected by the Egyptians around 2500 B.C. Ten centuries later, the Phoenicians used the first formal alphabet. These were all art forms and not printing, which is the reproduction of art forms in quantity.

Evidence of the first example of printing from movable type was discovered in 1908 by an Italian archaeologist on the island of Crete. He found a clay disc in the ruins of the palace of Phaistos in a stratification dated about 1500 B.C.

Printing from movable type appeared in China and Korea in the 11th century. In 1041, a Chinese, Pi-Sheng, developed type characters from hardened clay. They were not wholly successful. Type cast from metal in Korea was widely used in China and Japan, and by the middle 1200s type characters were being cast in bronze. The oldest text known was printed from such type in Korea in 1397 A.D.

Half a century later in 1440, probably unaware of the crude type developed in the Orient, Johannes Gutenberg introduced to the Western world his invention of printing with ink on paper, using movable cast metal type mounted on a converted wine press. Until Gutenberg's invention, all books were laboriously handwritten by scribes. Most historians credit his invention of printing as coinciding with the end of the Middle Ages and the beginning of the Renaissance. It is interesting to note that the invention of printing by Gutenberg has been declared the most important invention of the second millennium (1001-2000).

Paper and printing ink were not new when Gutenberg's type appeared. A Chinese, Ts'ai Lun, is credited with the invention of paper in 105 A.D. By the time Gutenberg was born, papermaking was a well-developed industry throughout the Western world with paper mills existing in Spain, France, Italy and Germany. The Chinese also led the world in making ink for printing. Wei Tang perfected an ink for block printing using lampblack in 400 A.D. Viscous or tacky inks, essential to printing, were already in use in Germany for block printing and for stamping titles on manuscript bookbindings. To Gutenberg we credit the invention of combining movable cast metal type, ink, paper and a press to produce printing that changed the world in the middle of this millennium.

TYPE BEGINNINGS

Our common typefaces are either imitations of early handwritten letters or represent a modification of early typefaces which, in turn, were modeled after the lettering in manuscript books.

The standard *Roman* lower-case letters and capital letters assumed their current form about 1470 in a face cut by Nicolas Jenson. While Jenson, a Frenchman, learned printing in Germany, he did his first printing in Venice, Italy. The letters inscribed in manuscript books by Venetian monks were Jenson's models. His types served as a pattern for later faces. Jenson was not the first to use Roman letters, but he must receive credit for developing a beautiful face upon which no later designer has been able to improve significantly. Typefaces similar to Jenson's models are often called Venetian types.

The first books in Europe were printed in block-letter or *gothic* type. They were designed to imitate the style of letter used by religious scribes living in the vicinity of Mainz, Germany, where Gutenberg began his printing activities. John Fust and Peter Schoeffer, who entered the printing field through business relations with Gutenberg, continued to use the gothic letter form. Thus, it became firmly established in Northern Europe.

To avoid confusion, it must be pointed out that the term "gothic," as used by some modern type founders to designate sans serif types, has no relation to gothic as a description of early typefaces. Gothic, as a term applied to architecture and other forms of art, designates the style characteristic of Northern and Western Europe from the 12th to the 16th century. It is in this sense that gothic is also applied to letter forms.

SETTING TYPE BY HAND

EARLY TYPEFACES

Roman Type

a

Gothic Type

a

ROMAN LETTER DEVELOPMENT

The manuscript hand of the Venetian scribes, which Nicolas Jenson followed as his model, developed apart from gothic lettering. It had evolved from Roman capital letters. In formal writing and inscriptions the early Romans used square capitals, with slight modifications, in the form of our upper-case alphabet. For correspondence and documents not requiring formal writing, large cursive or running capitals were used.

Many national styles in writing developed as learning was carried from Rome throughout the rest of the known world. The influence of the Roman characters might have been lost, however, had not Emperor Charlemagne taken an interest in the revival and spread of ancient learning. Charlemagne encouraged the establishment of a school at Tours by an English scholar named Alcuin. The calligraphy of this school became the model for the rest of Europe, and the introduction of what is now called lower case.

By the 10th century, the use of letter forms from which we derive our lower case was quite universal. However, these letters did not assume the fixed form with which we are familiar until they were cast in type by Jenson.

ITALICS AND DISPLAY TYPES

Practically all Roman typefaces in common use today have accompanying *italics*. This was not true of early Roman faces. Italics were first used to print small, compact books. Early books were large and cumbersome, and gothic type used in these books was large. When Roman type came into use, it was cast smaller than gothic, and letters and lines were fitted more closely. But even this economy in page size did not satisfy Aldus Manutius, a Venetian printer around the turn of the 15th century. Sensing a growing trend for cheaper books, he tried to meet the demand by cutting a font of type to imitate the informal hand-writing of his time. Aldus called this type *Chancery;* his Italian contemporaries called it *Aldine;* but in the rest of Europe, the face was called *italic*. This latter designation has continued in use to the present time.

Display types of today are difficult to trace historically. All were derived from hand-drawn letters. They may have been specifically drawn as a type-design or developed from a letter drawn for another purpose.

EARLY PRINTING IN ENGLAND

Early printing in England is interesting because it was through England that printing came to the American colonies. Printing was introduced in England about 1476 by William Caxton, who brought equipment from the Netherlands to establish a press at Westminster. Among the books issued from Caxton's press were Chaucer's *The Canterbury Tales, Fables of Aesop* and many other popular works.

The predecessor of the modern Oxford University Press was established in 1585. Since that date, the press has operated continuously, probably the longest period of any printing establishment in history.

Richard Pynson, who printed in England during the latter part of the 15th and early 16th centuries, is believed to have been the first to introduce Roman types in England. John Day, who began printing on his own account in 1546, was the first English designer of a Roman typeface.

TWO FAMOUS ENGLISH TYPE DESIGNERS

William Caslon, born in 1692 in Worcestershire, was destined to change the appearance of English printing through the design and casting of a new typeface. Not only is Caslon type still used, but his style of design is still consciously or unconsciously followed by many contemporary typographers. An axiom of printers with a type problem is, "When in doubt, use Caslon!" Although Caslon's letters are not perfect in themselves, a page of Caslon type produces a simple, pleasing and balanced effect.

The English printer and typographer, John Baskerville, born in 1706, is regarded by some students of printing history as the father of fine printing in England. Baskerville, after having accumulated a fair-sized fortune in other fields, established a paper mill, printing office and type foundry at Birmingham in 1750. Baskerville spent several years experimenting with designs for type. He also tried to improve the surface of sheets of paper by pressing them between hot plates after printing, and he mixed special inks which were used in producing his first book. Consequently, when he offered his first printed works to the public around 1757, they gained wide acclaim.

The types designed by Baskerville are usually considered to represent a halfway step between the *old-style* Roman letter which Caslon so clearly exemplified and the *modern* style of Roman letter which is best illustrated by the face developed by the Italian printer, Bodoni.

EARLY PRINTING IN AMERICA

Printing was used to promote colonization of the New World. On file in the New York Public Library is a copy of such a promotion piece dated 1609. It is entitled, "Offering Most Excellent Fruites by Planting in Virginia." One historian, observing the fact that 750 of the first 900 settlers in the Virginia colonies died during the first winter, marvels at the force of the printed word. It not only induced new settlers to come to the New World, but also influenced the 150 survivors to remain.

The first printing press in the New World was installed in Mexico in 1535. The first printing press in the British colonies made its appearance in Massachusetts in 1638, soon after the first settlers established themselves. The first piece printed on the new press was *The Freeman's Oath* in 1639. The *Bay Psalm Book,* 11 copies of which are still in existence, was produced in 1640. The press it was printed on was procured in England by Rev. Jose Glover who died on the voyage to America. His wife assumed responsibility for setting up the press in Cambridge. Stephen Daye, who had been indentured by Glover to operate the press, was placed in charge and, with his son, Matthew, continued its operation until 1647. The press became known as the Stephen Daye press.

In the meantime, Glover's widow was married again, this time to President Dunster of Harvard College. Upon her death, the press was moved to Harvard. In a sense, this represented the beginning of Harvard University Press, the oldest continuously-operated printing activity in America.

Printing did not make headway in the southern colonies to the extent that it did in the Massachusetts colony. By 1770, there were at least a dozen printers in Boston. By 1763, there was a press in operation in Georgia, the last of the 13 colonies to be settled. Printing came to Kentucky, Tennessee, Ohio and Michigan in the 1780s and 1790s. In 1808, printing had moved west of the Mississippi to St. Louis. Thus, as migration continued west, printing followed.

TWO PRINTER PATRIOTS

Benjamin Franklin was born in Boston in 1706. As a boy he learned printing in the shop of his brother. In 1723, he quarrelled with his brother and went to New York. Unable to find work, he continued on to Philadelphia where he worked for a French printer named Keimer.

At the suggestion of the governor of Pennsylvania, Sir William Keith, young Franklin went to England to buy a printing outfit. Money which he had been promised was not forthcoming, so for two years he worked in famous English printing plants, including that of William Watts. In 1726, he returned to Philadelphia. By 1732, he had his own printing office and became the publisher of the *Pennsylvania Gazette.* Among his publications, *Poor Richard's Almanack* became the most famous.

Throughout his life, Franklin was active in promoting printing. Although he disposed of his business in Philadelphia in 1748 to devote his time to literary, journalistic and civic activities, he assisted in the establishment and promotion of about 40 printing plants in the colonies. Franklin's high regard for his craft is revealed by the words with which he began his will: "I, Benjamin Franklin, Printer . . ."

Franklin is not the only printer of the Revolutionary Period who is celebrated as a great patriot. Isaiah Thomas, born in Massachusetts in 1744, was actively engaged in printing early in his life. In 1770, he began publication of the *Massachusetts Spy,* a newspaper in which he supported the cause of the patriots. He served during the Revolutionary War as printer for the Massachusetts House of Assembly. Following the war, he re-established his business, which had been destroyed. As a printer he prospered and became the leading publisher of books in the period following the Revolution. In 1810, he published a two-volume *History of Printing in America* which is the definitive source on colonial printing.

TYPE AND TYPECASTING MACHINES

For more than 400 years after the invention of printing, all type was set by hand. In the 19th century, men began to consider the possibility of creating typesetting machines. Numerous machines intended to replace hand composition were invented. The first of these was designed by an American, Dr. William Church, in 1822; others soon followed. While many of the first typesetting machines functioned satisfactorily, none were sufficiently practical for commercial operation until the invention of the Linotype by Ottmar Mergenthaler in 1886.

CHURCH'S TYPESETTING MACHINE

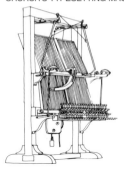

Of the various metal composing machines developed, only two kinds remained in use. These were the Linotype, Intertype and Ludlow machines which cast *slugs* (one-piece fully spaced lines); and the Monotype which cast *individual pieces of type* in justified lines. The Monotype was invented in 1887 by Tolbert Lanston of Washington, D.C. The Ludlow Typograph was suggested by Washington I. Ludlow in 1906 and later perfected by William A. Reade. Intertype, developed in 1911, used the Mergenthaler principle. Phototypesetting, introduced in 1949, started new developments followed by electronic typesetting in 1970, and the plain paper digital typesetter in 1985 which has almost completely replaced mechanical typesetting.

PLATEMAKING BEFORE PHOTOGRAPHY

The first illustrations in books were made from woodcuts. They were tooled out of wood blocks by hand, leaving raised surfaces. The earliest known book using woodcuts was printed by Albrecht Pfister in Bamberg, Germany about 1460.

Books printed between 1570 and 1770 were usually illustrated by copperplate engravings, resulting in a decline in the making of woodcuts. In 1770, however, a revival was started by Thomas Bewick of England who developed the technique of using a special engraving tool for cutting *across* the grain, instead of *with* the grain. Later, woodcuts were used only to give an "artistic touch" to certain types of printed pieces.

Engraved copper intaglio plates, the forerunner of steel engravings and gravure, were first used in France and Italy around 1476. Copper engraving offered competition to woodcuts in England about 1545, and in France about 1569. Copperplate engraving has continued to be used through the years and is still used for some invitations and announcements.

PHOTOGRAPHY AND PHOTOMECHANICS

Photography for graphic arts involved the photographic processes and techniques used to reproduce illustrations and art subjects. Photomechanics or photoplatemaking, like photoengraving, photolithography, photogelatin, photogravure, etc., was the means of using light sensitive coatings and halftone and line films to make plates and cylinders for printing. The invention and use of photography and photomechanics completed the mechanization of the printing process, made illustrations practical and economical to produce and reproduce, and fostered the phenomenal growth of advertising, periodical, book and commercial printing.

In 1839, the year Daguerre invented photography, Ponton discovered the use of potassium bichromate as a sensitizer, and in 1852, Fox Talbot used it to sensitize gelatin and produced a halftone engraving by laying a screen of fine gauze between the coated metal and a negative of the original picture. This is the first known use of the screen principle which created the "dot pattern" as it is known today. In 1855, Poitevin invented photolithography based on bichromated albumin. Photoengraving developed rapidly in America, and by 1871, it became practicable for letterpress printing. By 1880, photoengraved prints were replacing woodcuts as illustrations in books, magazines and newspapers.

The first commercial halftone screen was produced in 1883 by Max and Louis Levy of Philadelphia. Two years later, Frederick Ives improved on their technique by developing the earliest version of the glass crossline screen. Although the first halftones were black and white, the application of halftones to color process printing was not far behind. The first color process work was printed successfully in 1893. Photography is gradually being replaced by digital imaging.

DEVELOPMENT OF THE PRINTING PRESS

While hand composition did not change much through the years, a number of changes were made in transferring the impression to paper. The Stephen Daye press (1638) was similar to the Gutenberg press. Benjamin Franklin worked on a wooden-frame press in the printing office of William Watts in London which was an improvement over the Stephen Daye press. This press used a torsion screw for making the impression and was provided with a clever mechanical arrangement devised to provide the proper pressure on the form.

Further changes in press construction came about slowly until the first all-metal press was built by the Earl of Stanhope early

STEPHEN DAYE PRESS

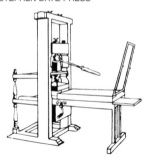

in the 19th century. This press still used a screw device, but less exertion was required to force the impression on the sheet. Application of the principle of the lever to the iron press resulted in several presses which came into common use. Among these were the Columbian press, the Albion press, and the Washington press. The Washington press became popular in the United States, and by 1900 more than 6,000 had been sold. The Albion press was equally popular in England.

The idea of the printing press, as conceived by Gutenberg, reached its highest development in the Washington and Albion presses. During the Industrial Revolution, in the 1800s, new designs for printing presses were developed. Two of these were (1) the job, or platen, press, and (2) the cylinder press. The job press is the direct descendant of a machine perfected in 1858 by George P. Gordon of New York. In this machine the platen and form are turned on edge. In others, both the platen and the bed move with a sort of clamshell action.

The first successful cylinder press was the steam-powered cylinder press built in London under the supervision of a German named Frederick König. It was used for printing *The London Times* in 1814 and was capable of producing 1,100 sheets per hour. A rotating cylinder was used to press the paper against a flat type bed. Shortly after König's press was placed in operation in 1814, D. Napier, an Englishman, invented a press using grippers for picking up the sheet from the paper table and holding it while the sheet received the impression. While many other improvements were added, the flatbed cylinder press, except the vertical press, is obsolete.

In the United States, Richard Hoe perfected the first rotary press in 1847, with the type actually carried on the cylinder. Early models produced 2,000 impressions per hour per cylinder.

The first web press was also developed by an American,

HOE'S TEN-CYLINDER PRESS

William Bullock, in 1856. A similar press was patented 10 years later in London. These early presses delivered 15,000 signatures per hour and printed both sides. A device for folding the papers as they came from the press was added in 1875.

Since that time, newspaper presses have been developed to a high state of efficiency which, by duplicating plates and units, has allowed newspapers to be printed and delivered at the rate of 160,000 per hour. Web presses are used almost exclusively by gravure and have been a dominant factor in lithographic printing since 1954.

HISTORY OF LITHOGRAPHY

The basic principle of lithography, which means literally "writing on stone," is actually based on the principle that grease, or oil, and water do not mix. It was discovered by Alois Senefelder of Munich about 1798. Working on a highly porous stone, he sketched his design with a greasy substance which adhered to the stone. He then wet the entire surface with a mixture of gum arabic and water. It wet the blank or non-image areas, but the greasy design repelled it. An ink made of soap, wax, oil and lampblack, was rolled on the stone. This greasy substance coated the design but did not spread over the moist blank area. The design transferred to the paper when the paper was pressed against the surface of the stone.

Artists soon used lithography to make reproductions of the works of old masters and, in time, it became a valuable medium for their own original works. Currier and Ives popularized lithography in the middle of the 19th century and it soon became a more practical and faster method of printing illustrations.

SENEFELDER'S PRESS

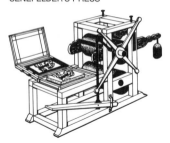

The first steam press for lithography was invented in France in 1850, and was introduced in the U.S. by R. Hoe in 1868. The first use of the *offset principle* by lithography was for metal decorating about 1875. Lithographic stones were used for the image. A blanket-covered cylinder received the image from the stone and transferred it to the metal. Direct rotary presses for lithography were introduced in the 1890s using grained zinc and aluminum metal plates to which images were hand transferred from stones using starch-coated transfer sheets. These plates had difficulty printing on rough surface papers.

In 1906, the first "offset" rotary press began printing sheets in Nutley, N.J., an invention of Ira A. Rubel, a paper manufacturer. Actually, the discovery was accidental. An impression was unintentionally printed from a plate cylinder directly onto the rubber blanket on the impression cylinder. Immediately afterward, when a sheet of paper was run through the press, a sharp image was printed on it from the impression which had been "offset" on the rubber blanket. A.F. Harris had a similar experience which encouraged him to develop an offset press for the Harris Automatic Press Company of Niles, Ohio, in the same year, 1906.

HISTORY OF PRINTING PAPERS

As already stated, papermaking was invented in China more than 1,800 years ago. By 1200 A.D., paper was being made in Spain, and 200 years later the art was well established throughout Europe. The first paper mill in England was built in 1494. In the American colonies, paper was first manufactured commercially in 1690, in a mill near Philadelphia owned by William Rittenhouse. Paper originally was made for uses other than printing, but soon after the invention of the printing press it became one of the most important uses of papers.

Ancient papers were made almost entirely from rags and were produced with crude hand-operated devices. Most papers currently used in printing are manufactured from wood pulp. A few writing and ledger papers are still made from rags, but even these contain a percentage of wood pulp.

The machine for producing a continuous web of paper using a wire mesh screen to form the paper was invented by a Frenchman, Louis Robert, in 1798. His invention was financed and developed by an English family, the Fourdriniers, and even today, a papermaking machine may be referred to as a "Fourdrinier." An important advance in paper machines has been the introduction of the twin wire machine which reduces the effects of two sidedness on printing.

Paper manufacturing from mechanical groundwood pulp was introduced in 1840. Production of cellulose, or wood fiber, by chemical methods, using caustic soda, was perfected in 1854. Bisulphite of lime came into use about 1866. More recent developments are thermomechanical pulping, alkaline sizing and the use of more recycled fibers.

HISTORY OF PRINTING INKS

As previously stated, ink was already in use for printing from wood blocks at the time Gutenberg developed his movable type system. Actually, the origin of printing ink is shrouded in mystery. It developed from writing ink which was used by the Egyptians and Chinese as early as 2600 B.C. These early inks consisted of lampblack or soot mixed with animal glue or vegetable oils. Inkmaking became a highly developed art among the Chinese as they introduced earth colors and printed from hand-cut blocks in the 11th century — 400 years before Gutenberg.

In the early days, the printer made his own inks with lampblack and boiled linseed oil which he cooked according to a "secret" formula. Much of the success of Gutenberg's invention of printing is due to the special ink he developed for transfer to and from the cast metal type. Inkmaking became a commercial process in the 17th century. The first ink factory was established in America in 1742. Little color was used until the discovery of coal tar dyes in the middle of the 19th century. Linseed oil was the main vehicle in printing inks until the mid-1930s when new vehicles and heat-set inks were introduced for letterpress magazine printing in the U.S. UV (ultraviolet) and EB (electron beam) curing vehicles for inks and coatings were introduced in the 1970s. More recent developments in inks have been water-based inks for gravure and flexography, soybean inks for lithography and special inks for waterless printing. The latest developments in printing inks for lithography are single fluid, or emulsion type inks.

DIGITAL IMAGING AND PRINTING

Digital Imaging The age of electronics and computers has changed the way printed products are created and produced. The first printing production operation to be affected was typesetting with the introduction of the Fotosetter in 1949 and the Photon in 1954. In 1950 the PDI Electronic Scanner was introduced to perform color separations. Neither technology advanced much until the 1970s when the Video Display Terminal (VDT) and Optical Character Recognition (OCR) were introduced to improve electronic typesetting, and Electronic Dot Generation (EDG) and digital magnification expanded the capabilities of electronic scanners.

The digital revolution in typesetting occurred in 1985 with the introduction of the Plain Paper Typesetter which became the Imagesetter in 1988. Digital imaging in prepress started to expand in 1979 with the introduction of color electronic prepress systems (CEPS). CEPS were expensive, device-dependent systems with little or no interaction between systems. In 1988 the imagesetter and raster image processor (RIP) fostered the development of device independent prepress systems known as Desktop Publishing which displaced the device dependent CEPS and eventually replaced conventional prepress systems.

Digital imaging for platemaking started with laser platemaking about 1975; laser engraved cylinders for flexography and engraving assists for gravure in the 1980s; computer-to-film plates in the 1980s; computer-to-metal plates in 1991; thermal laser ablation no-processing computer-to-on-press plates in 1993; and computer-to-thermal processless plates starting in 1997.

Digital Printing Conventional printing uses plates which contain the images to be reproduced in quantity on a mechanical press that feeds inks to the plates and exerts heavy pressures to transfer the inked images to paper or other substrate. Digital printers, on the other hand, are like copiers. They are plateless systems. Each cycle of the printer transfers a fresh image to the substrate. It can be the same or different than the previous image. This feature makes it possible to print variable information from print to print which conventional printing cannot do. Digital printing is used for short run, on-demand printing but the speed is much slower than plate processes because of the tremendous amount of memory required to repeat or change the image in each cycle.

There are a number of variations of digital printers that use many different technologies. Digital printing started in 1970 with the introduction of ink-jet printing, followed by electrophotographic (EP) laser printers in 1978; color electronic laser printers starting in 1990; and two EP color printing presses introduced in 1993. Other digital printing technologies are Ion or Electron Charge Deposition, Magnetography, Thermal Transfer, Thermal Transfer Dye Sublimation, and Electro-Coagulation printing.

THE PRESENT

Digital imaging has already transformed prepress. Photography has been almost completely replaced by digital imaging systems. Dry processless films are in use. Digital cameras are decreasing the need for scanners. Prepress for all printing processes has been almost completely converted to device independent desktop publishing hardware and software. Most color proofing has shifted from analog (film based) to digital proofing. FM (stochastic) screening favors continuous tone over halftone digital proofing. Imagesetters have advanced from film polyester-based plates (Imposetters), to fully-imposed metal plates (Platesetters).

Computer-to-plate technology is gradually being adopted. Digital presses are finding niche markets for short run, on-demand and variable information printing. Press makereadies are being shortened by computer-to-plates, automatic setting of ink fountain keys by digital data, automatic plate changers, and increased use of waterless printing. Offset press speeds of 3,000 feet/minute (fpm) have been reached by the use of gapless blankets and plate sleeves on web offset presses.

Computers are used everywhere in the printing process. Color is measured and controlled on- and off-press by densitometers and spectrophotometers. Paper is transported to the press and signatures from the press to the bindery automatically by intelligent robots. Finishing is done on- and off-line.

On-demand and variable information printing by digital printers is beginning to change the basic concept of printing in some printing markets from *print and distribute* to *distribute digital files and print on location.*

THE FUTURE

Continued developments in and applications of digital imaging and printing will characterize the future of printing. The use of portable document formats (PDF) to move a digital file to an output service will help solve the problems of PostScript® errors and reduce the need for preflighting of digital files. Interactive remote digital color proofing, using FM screening and either dye sublimation or ink-jet continuous tone imaging systems will simplify digital workflows and speed the conversion to computer-to-systems for conventional printing. Computer-to-plate and computer-to-plate-on-press technologies will advance with the availability of no-process thermal plates. Waterless printing will expand in use with the availability of other sources of printing plates. Printing will enter a filmless era and digital printing will expand to many new markets using low cost ink-jet and thermal transfer printers as well as electrophotographic copier/printers.

Offset presses are coming close to stabilization and predictability with automatic control of inking, plate and press temperatures, dampening, and color and tone reproduction at speeds over 3,000 feet per minute (fpm). Binding and finishing will continue to be automated and alternate means of distribution will be developed.

Lithography will continue to be the dominant printing process well into the 21st century. Flexography with water base and UV inks will continue to gain in use. Gravure, despite developments in direct digital imaging will lose market share as press runs decrease in length. Letterpress is being revived in some market areas by the use of UV inks and narrow web presses. By the year 2010, about 80 percent of the printed products in the U.S. will be produced by conventional printing processes; and 20 percent representing short-run, on-demand, variable information, and other specialty printing products will be produced by digital printers.

The Internet will be used to produce many products that are presently produced by printers. Progressive printers who recognize this new competition to printing as an opportunity will prosper. Those who don't will either go out of business or be absorbed by the progressive printers. It has been estimated that by 2010 there will be 7,000 less printers in the U.S. than in 2000. Consolidations have mushroomed in the past several years and will continue to increase.

While printing is sure to change more in the next 20 years than it has in the more than 550 years since Gutenberg, the printed word on a paper-like substrate will survive and continue to flourish well into the 21st century. It is certain to change in the way it is composed and produced, but it will be around for many years for people to handle, carry, fold, read, admire and enjoy.

Introduction To The Printing Process

A number of printing processes are used to produce the many different types of printed products. Each printing process has three main operations, or steps: *Prepress, Press,* and *Postpress.* The prepress and postpress steps are very similar for all printing processes. The prepress step involves the preparation of the copy or information for printing. Before the introduction of electronics and computers in printing, much of prepress operations were manual using handset type and/or typesetting machines, process cameras for imaging functions, film processing, manual color correction, film assembly, and page and signature layout. *Digital imaging* with computers and software has almost completely replaced all manual operations, but prepress still is similar for all printing processes. Postpress is also similar for all processes, but it has not been affected as much by computerization.

The press or printing step uses presses that are different for each printing process. There are two major classifications of printing processes: (1) *plate,* pressure or impact processes like conventional offset lithography, letterpress, flexography, gravure and screen printing; and (2) *plateless* or pressureless processes like electrophotography, ink-jet, ion or electron charge deposition, magnetography, thermal transfer printing, thermal dye sublimation and electro-coagulation.

Plate printing processes use mechanical printing presses to exert heavy pressures needed to transfer ink to paper, and are the processes used to support markets like magazines, newspapers, books, packaging, and many other printed products.

Plateless printing systems are the processes used for copying and digital printing. They use digital imaging and copier-like printers that produce an image during each cycle of the printing device. The image can be the same or can be changed from cycle to cycle. This feature enables digital printers to print variable information from cycle to cycle such as coding, addressing and personalizing documents. The printing speed, however, is much slower than plate processes as the amount of computer memory required for repeating and/or changing information in each cycle is enormous. Besides productivity, these processes have cost limitations. In general, costs are the quantity of prints times the unit cost. Therefore they are used mainly for short runs, on-demand, and/or variable information printing.

CONVENTIONAL PRINTING PROCESSES

All printing processes are concerned with two kinds of areas on the final print: (1) *Image* or printing areas, and (2) *non-image* or non-printing areas. After the copy or information has been prepared for reproduction (the prepress step), each printing process has definitive means of separating the image from the non-image areas. Conventional printing has four types of printing processes: (1) *Planographic* in which the printing and non-printing areas are on the same plane surface and the difference between them is maintained chemically or by physical properties. Examples are *offset lithography, collotype* and *screenless printing;* (2) *Relief* in which the printing areas are on a plane surface and the non-printing areas are below the surface. Examples are *letterpress* and *flexography;* (3) *Intaglio* in which the non-printing areas are on a plane surface and the printing areas are etched or engraved below the surface. Examples are *gravure* and *steel-die engraving;* (4) *Porous* in which the printing areas are on fine mesh screens through which ink can penetrate, and the non-image areas are a stencil over the screen to block the flow of ink in those areas. Examples are *screen printing* and *stencil duplicator.* Images by each printing process have distinguishing features illustrated below.

Planographic – Offset Lithography

This is the major plate printing process. It uses thin metal plates with the image and non-image areas on the same plane. There are two basic differences between offset lithography and other processes: (1) it is based on the fact that *oil and water do not mix*, and (2) it uses the *offset principle* in which ink is offset from the plate to a rubber blanket on an intermediate cylinder, and from the blanket to the paper on an impression cylinder.

On a lithographic printing plate, the printing areas are oil or ink-receptive and water-repellent, and the non-printing areas are water receptive and ink-repellent. When the plate mounted on the plate cylinder of the offset press is rotated, it comes into contact with rollers wet by a water or dampening solution and rollers wet by ink. The dampening solution wets the non-printing areas of the plate and prevents the ink from wetting these areas. The ink wets the image areas which are transferred to the blanket cylinder. The inked image is transferred to the paper as the

HOW TO RECOGNIZE THE PRINTING PROCESSES

LITHOGRAPHY	FLEXOGRAPHY	GRAVURE	SCREEN
(Smooth Edges)	(Ring of Ink)	(Serrated Edges)	PRINTING
	Rotary Letterpress		(Screen Edges)

paper passes between the blanket cylinder and the impression cylinder. Letterpress and gravure can also be printed by the offset principle. Because practically all lithography is printed in this way, the term *offset* has become synonymous with lithography.

SINGLE COLOR OFFSET PRESS

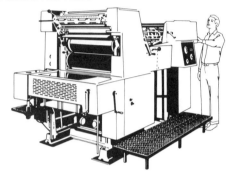

A main advantage of the offset principle is that the resilient rubber blanket produces a clearer impression on a wide variety of paper surfaces and other materials with both rough and smooth textures. Also it increases plate life and reduces press makeready. Offset printing produces a smooth print without embossing, an ink ring, or serrated edges which are characteristic of letterpress, flexography, gravure and screen printing.

Offset lithography has equipment for short, medium and long runs. Both sheetfed and web presses are used. Sheetfed lithography is used for printing advertising, books, catalogs, greeting cards, posters, labels, packaging, folding boxes, decalcomanias, coupons, trading stamps and art reproductions. Also many sheetfed presses can *perfect* (print both sides of the paper) in one pass through the press. Web offset which prints on rolls of paper is used for printing business forms, newspapers, inserts, advertising literature, long-run catalogs, books, encyclopedias and magazines.

Relief – Letterpress

This is the method of printing Gutenberg invented in 1440 and has been used for job and commercial printing. It is a *relief* method of printing that can print from cast metal type, molded duplicate plates (almost obsolete) or photopolymer plates on which the image or printing areas are raised above the nonprinting areas. Viscous oil-base and UV inks are used. The ink rollers come in contact with the raised areas only, and the inked image is transferred directly to the paper.

Commercial letterpress has declined in use because too

much time was consumed in *makeready* (building up of the press form so both the light and heavy areas print with the correct impression). Makeready is a manual process that is very skill- and time-intensive and therefore was very expensive. It is a main reason letterpress declined in use.

Four types of presses were used: platen, flatbed cylinder, rotary and belt *(see page 139)*. Letterpress is making a comeback in specialty printing using photopolymer plates and UV inks on narrow web presses. The distinctive feature for recognizing letterpress and flexography is a heavier edge of ink around each letter (seen with a magnifying glass). Sometimes a slight embossing appears on the reverse side of the paper. The letterpress image is usually sharp and crisp, but grainy.

Relief – Flexography

Flexography is a form of rotary web relief printing like letterpress but using flexible rubber or resilient photopolymer relief plates, and fast-drying low viscosity solvent, water-based, or UV inks fed from an anilox inking system. Flexographic presses are web-fed in three types: *stack, in-line* and *central impression cylinder* presses. Almost anything that can go through a web press can be printed by flexography. Products printed by flexography range from decorated toilet tissue to bags, pressure sensitive labels, corrugated board and materials such as foil, hard-calendered papers, cellophane, polyethylene and other plastic films. It is well suited for printing large areas of solid color with high gloss and brilliance.

PRINCIPLE OF FLEXOGRAPHIC STACK PRESS

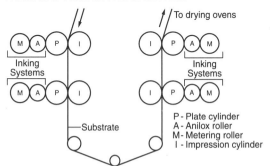

P - Plate cylinder
A - Anilox roller
M - Metering roller
I - Impression cylinder

The growth of flexography parallels the expansion of the packaging industry, the development of the central impression cylinder press, laser engraved ceramic anilox ink metering systems, reverse angle doctor blades, photopolymer plates and UV inks. Halftones as fine as 150 lines per inch can be printed on flexible films. It has also gained prominence in the printing of business forms, books, newspapers, labels, folding cartons and corrugated boxes, as well as many specialty items such as gift wraps and shopping bags. The improved quality of photopolymer plates controlled inking of advanced ink metering systems with reverse angle doctor blades, and water-based and UV inks have favored flexography to replace letterpress, gravure and lithography in some printing markets.

Intaglio – Gravure

Gravure image areas consist of cells or wells etched or engraved into a copper cylinder, and the unetched surface of the cylinder represents the non-printing areas. The image cylinder rotates in a bath of ink. The excess is wiped off the surface by a flexible steel *doctor blade*. The ink remaining in the thousands of recessed cells forms the image by direct transfer to the paper as it passes between the plate cylinder and the impression cylinder. Three types of processes are used for making gravure printing cylinders: (1) *chemical etching* that produces cells of the same size or area with varying depths or cells with varying area and depth; (2) *electromechanical engraving (EME)* which produces cylinders with cells that vary in area and depth; and (3) *direct digital laser etching* process which uses a special alloy and laser to produce cells of varying area and depth at speeds of about 70,000 cells per second which are from 10 to 15 times higher than EME.

Gravure printing produces excellent reproductions of pictures, but slightly ragged type. The high cylinder-making cost usually

GRAVURE PRESS

limits its use to long runs. The use of halftone and filmless gravure has reduced these costs and made gravure competitive in shorter run markets. Gravure is used for long runs of newspaper supplements, magazines, catalogs and packaging.

Porous – Screen Printing

Formerly known as silk screen, this method employs a *porous* screen of fine silk, nylon, dacron or stainless steel mounted on a frame. A stencil is produced, either manually or photomechanically, in which the non-printing areas are protected by the stencil. Printing is on paper or other substrate under the screen by applying ink with a paint-like consistency, spreading and forcing it through the fine mesh openings with a rubber squeegee.

PRINCIPLE OF SCREEN PRINTING

Printed Image | Screen | Squeegee

The production rate, formerly limited by the drying time of the ink, has been greatly increased through the development of automatic presses, improved dryers and UV inks. New rotary screen presses speed up production considerably because they allow continuous operation. Screen printing usually can be recognized by the thick layer of ink and sometimes by the texture of the screen on the printing.

OTHER CONVENTIONAL PRINTING PROCESSES

Copying

Copying and duplicating are also called *reprography.* For fewer than 10 copies, the copier offers the fastest, most economical method of duplication. Above 10 copies, high-speed copiers and/or duplicators are used. Reprography is used extensively by in-house printing departments and quick printing shops.

The only method for making copies of documents before 1940 was the *photostat,* which was cumbersome, time-consuming and expensive. Since then, a number of photocopying systems were developed like the diffusion transfer processes in Europe. These processes were displaced by the 914 Xerox Copier in 1960. Since then, electrophotography has dominated the copier market.

Electrophotography, also called xerography, is based on electrostatic transfer of toner to and from a charged photoconductor surface. Both plain paper and special coated-paper copiers use electrophotographic coatings such as selenium, cadmium sulfide, zinc oxide or organic photoconductors to produce the images in the copier. These coatings are charged with a corona discharge in the dark and lose the charge on standing or when exposed to light such as that reflected from the white areas of an original or by a laser. The image areas which remain charged are developed with an oppositely charged dry powder or liquid toner and are transferred from the electrophotographic surface onto plain paper using electrostatic attraction. The electrophotographic coating is cleaned and can be re-imaged many times. Color copiers use similar principles.

Sophisticated copiers and printers can have many special features that make them complete publishing systems with binding and finishing. These systems can be enhanced with interfaces to digital information and be used as digital presses for on-demand and short-run printing.

Duplicating

Before the introduction of digital presses, when single color copies of documents in quantities from 50 to 10,000 were needed, the most economical printing method was the offset duplicator.

Offset duplicator is a small offset press in sizes up to 12x18 inches. These presses are described on page 137.

OFFSET DUPLICATOR

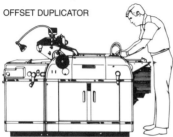

Copier/duplicators automate duplicating by combining a *copier* and a *duplicator*. In this system an electrophotographic copier makes the plate, and a transport device automatically mounts the plate on the duplicator press where it is treated with a chemical solution; the press prints the number of copies programmed in the counter, rejects the old plate, and either stops or inserts a new plate and continues to print.

Stencil and spirit duplicators These two less expensive and less sophisticated systems for duplicating have been in use for many years. The *stencil duplicator,* or mimeograph, which works by forcing ink through a porous stencil usually prepared on a typewriter, or by spark discharge, produces copies of acceptable quality on plain paper. These have been digitized so that pages can be composed using a digital layout tablet and a light pen. Also the duplicator can be imaged by a desktop publishing system using a special digital interface.

The *spirit duplicator* is unique in that the master can be made to print more than one color at the same time using special carbon papers which contain soluble dyed resins of different colors. A special printing paper is wet with a fast-drying solvent which, when brought in contact with the spirit master, softens and tackifies the resin so that a small amount of coloring dye is transferred to the paper. The master can be printed, and reprinted, until all the dye in the ink is used up, which is usually about 100 prints.

DIGITAL PRINTING

Digital printing is a combination of *digital imaging* and *digital press*. Digital imaging has already almost completely replaced *conventional prepress*. In doing so it has created an intermediate type of printing process with *digital prepress* and *conventional press*. Digital imaging has made possible the filmless imaging of (1) printing plates for conventional presses and (2) image carriers for digital presses.

Filmless digital imaging processes There are three types of filmless digital imaging processes that create images directly on printing plates for conventional printing, or directly on plateless presses. These are called *computer-to-,* or *direct-to-* printing processes (to avoid repetition this book will use the term *computer-to-* throughout). The three types of filmless digital printing processes are (1) *computer-to-plate* (CTPe), (2) *computer-to-plate-on-press* (CTPs) and (3) *computer-to-print* (CTPt). All the categories use essentially the same front end or prepress digital files of page composition software, or page description language and a raster image processor (RIP) to drive the lasers in the imaging devices. Categories (1) and (2) use the digital files to produce plates that print on conventional or modified printing presses; category (3) uses the digital files to produce the images on existing plateless processes or new digital printing systems with the same features as plateless processes.

CONVENTIONAL/DIGITAL PRINTING SYSTEMS

Computer-to-Plate (CTPe) Systems

Computer-to-plate or image cylinder technology was first used commercially by gravure and flexography in the 1980s followed by lithography in 1990.

Gravure All publication gravure is produced by computer-to-image cylinders on EME machines using interfaces to digital imaging systems. Also a lasergravure process has been introduced capable of etching over 70,000 cells/second on a special new alloy. Copper is not suitable for lasergravure because it reflects laser beams. The speed of etching of the new process is 10 to 15 times faster than EME.

Lithography Eight new plate technologies have been developed, modified, or are in testing to implement computer-to-plate technology for lithography. They are silver halide, high speed dye-sensitized photopolymers, hybrid mask double coated, thermal laser ablation, laser ablation transfer, thermal crosslinking, ink-jet and new no-process plate technologies.

Computer-to-Plate-on-Press (CTPs) Systems

The first CTPs technology was a dry *spark discharge* plate in 1991 which required no liquid processing so it could be imaged on-press and could be run without dampeners. The imaging systems were mounted in place of the dampening systems. The spark discharge technology was replaced by thermal laser ablation in 1993 which produced much sharper dots and higher image quality. Other systems in development use gapless sleeves, on-press coating, imaging and printing, followed by erasing and repeating the operations.

DIGITAL PRINTING SYSTEMS

Computer-to-Print (CTPt) Systems

Most plateless printers can be adapted to accept digital files and become computer-to-print (CTPt) digital printing systems. They use electrophotographic photoconductor, ink-jet, ion or electron charge deposition, magnetography, thermal transfer, thermal dye sublimation and electro-coagulation technologies.

Electrophotography (EP) In 1993 two new (CTPt) systems were introduced, one using liquid toner EP photoconductor technology on a sheetfed offset press design; the other using dry toner EP technology on a web press design. These systems are developing markets for short-run color, on-demand and variable information printing. Other printing systems using EP engines are in use. One is a press design like the first two, and the others are clusters of EP color copier engines.

DIGITAL WEB PRESS

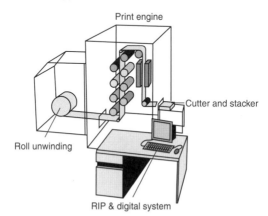

Most color copiers can be converted to digital color printers by using appropriate *color servers*. All laser printers which started as intelligent copiers in the late 1970s can also be used as digital color printing systems. These can be arranged into cluster systems, too. There are also several *large format* EP printing systems used for producing displays, murals, outdoor exhibits, posters, and paneled billboards.

Ink-jet systems use jets of ink droplets driven by digital signals to print the same or variable information directly on paper without a press-like device. The first plateless digital printing system was ink-jet introduced in 1970. Two types of ink-jet systems are in use: (1) continuous jets and (2) thermal jets.

(1) Continuous jets are used for digital color proofing, color printers and variable information on-press color printing systems at speeds up to 220 feet/minute. (Single color ink-jet printing has been done at 1,000 fpm on a newspaper press.)

(2) Thermal jets are divided into *Bubblejet* and *Solid Ink/ Phase Change* digital printing systems.

Ink-jet systems are gaining in use, especially since new inks have been developed with light- and water-fastness properties.

Ion or electron charge deposition is imaged by passing digital data through an electron cartridge that generates negative charges which produce a charged image on a heated dielectric drum. The visible image is created with a magnetic toner which is transferred to the substrate by cold fusion.

Magnetography is similar to electron deposition except that a magnetic drum, a magnetic charge and special magnetic toners are used to produce the printed images.

Thermal transfer printers use digital data to drive thermal print-heads in press-like machines that melt spots of dry thermo-plastic ink on a donor ink ribbon and transfer them to a receiver to produce color labels, logos, wiring diagrams, bar codes and other similar products.

Thermal dye sublimation printers are like thermal transfer printers except the inks on the donor ribbons are replaced by sublimable dyes. The thermal head converts the dyes to gas spots that condense on the receiver. The printers are used for digital color proofing and color printing.

Electro-coagulation printing is a new digital printing technology. It uses a pigmented water-base ink sprayed on a release coating on a metallic press cylinder. The image is produced by a digital file-controlled printhead that uses the cylinder as a positive electrode to coagulate the ink imagewise on the release coating. This transfers the coagulated ink image to the paper in printing.

All these printing systems are described in more detail and some are illustrated in the sections on *Digitally Imaged Presses (page 145)* and *Digital Presses (page 147)*.

Prepress

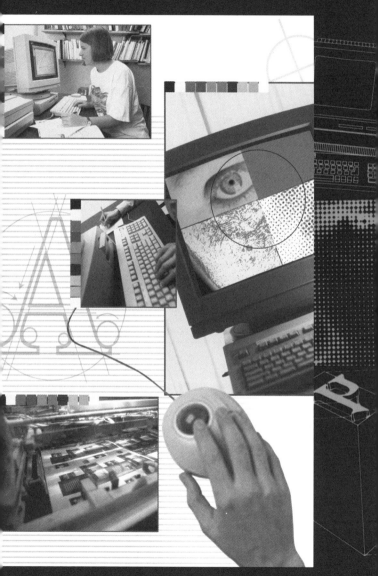

INTERNATIONAL PAPER

TYPE AND TYPOGRAPHIC IMAGING

Printing is the reproduction in quantity of words and pictures on a page or document. *Prepress* is the first category of operations to accomplish the reproductions by printing, which is followed by *Press,* and *Postpress.* Prepress is the series of operations involved in the preparation and assembly of all copy elements ready for printing on a printing press or digital printer. Copy elements include all materials to be reproduced in text or picture form. This section deals with text, or type, how it has been set in the past and how digital imaging has affected its handling.

Type has a number of different characteristics and styles which are divided into numerous classifications. It is important to know all these type fundamentals before becoming familiar with the various methods of typesetting. While digital typesetting has uprooted the basic methods of typesetting it still must adhere to the type fundamentals and classifications of the typefaces used for reproduction.

TYPE FUNDAMENTALS

Typefaces are usually available in 6- to 72-point, with a complete *font* in each size. A font is defined as a complete assortment of any one size and style of type containing all the characters for setting ordinary composition.

Capital letters are called *upper case* and small letters, *lower case.* When old-time compositors set type by hand, they placed the case with the capital letters above the case with the small letters, thus the nomenclature.

In lower-case letters the upper stroke (as in the letter "b") is called the *ascender,* and the downward stroke (as in "p") is known as the *descender.* The short crossline at the end of the main strokes is called the *serif.* Typefaces without serifs are called *sans serif.* The body or *x-height* makes up the greatest portion of a letter.

THREE MAIN CHARACTERISTICS OF TYPE

Ascender	
x-Height	**bop**
Descender	

Type size cannot be measured from the top of an ascending letter to the bottom of a descending letter. The face of any letter is not the full point size. For example, the face of a 36-point letter may measure only 30 points.

Corresponding letters in the same size type may vary in height. We say that the face is either small on body (small x-height) or large on body (large x-height). For example, the following are all 24-point lower case "h":

h **h** h **h** h **h** h **h** **h** h

To most people, many typefaces look alike; even an expert must look carefully to differentiate among them. While there is no short cut in learning how to identify typefaces, careful study of a few key letters helps. For example, the lower case "g" is one of the most distinctive letters. The elements are: the top loop, the lower loop, the hook off the top loop and the element joining the two loops. By studying the size, shape and position of these elements, the typeface identity can be determined more easily. Other distinctive letters are "p," "a," "e" and "t".

HELVETICA MEDIUM IN 11 DIFFERENT SIZES

6	ABCDEFGHIJKLMNOPQRSTUVWXYZABCDEFG
8	ABCDEFGHIJKLMNOPQRSTUVWXYZ
10	ABCDEFGHIJKLMNOPQRSTU
12	ABCDEFGHIJKLMNOPQR
14	ABCDEFGHIJKLMNOP
18	ABCDEFGHIJKLM
24	ABCDEFGHI
30	ABCDEFG
42	ABCDE
48	ABCD
60	ABC

The selection of typefaces is a very critical operation. The entire appearance of a printed piece can be affected. Many characteristics—masculinity, femininity, delicacy, formality, etc.—can be suggested by the typeface used. The guidance of a qualified designer, printer or typographer in selecting the proper typeface is indispensable.

Above all, remember type was designed to be read easily! Both the selection of the typeface and the size to be used must be considered. Use italics with care! Their primary purpose is for emphasis, not to be read in a mass.

CLASSIFICATION OF TYPEFACES

There are many approaches to type classification, none of which is precise. The following, however, is a useful breakdown covering a wide variety of typefaces:

Oldstyle, patterned after letter forms used on classical Roman inscriptions, looks better in mass than when examined letter by letter. The letters have high readability because they are open, wide and round, with pointed serifs that make a pleasing contrast between the heavy and light strokes. Garamond and Caslon are examples.

Modern refers not to a time period, but to a style of type designed almost 200 years ago. These types have a much greater degree of mechanical perfection than Oldstyle faces and are distinguished by extreme contrast between thick and thin strokes, with thin, squared-off serifs. Bodoni and Times Roman are examples.

Square serif is a contemporary type style used mainly for display, headlines and small amounts of reading matter. The letters have square or blocked serifs and more or less uniform strokes, and the face is even in texture and weight with very little contrast. Clarendon, Stymie and Lubalin Graph are examples.

Sans serif enjoys great popularity because of its simplicity of design. The letters have no serifs, and the face is generally even in overall weight with very little contrast between thick and thin strokes. Helvetica, News Gothic and Futura are examples. (The *Pocket Pal* is set in Helvetica.)

Script, designed to simulate handwriting, is used mostly for special effects, formal invitations and announcements. There are no serifs or extreme contrasts between the thick and thin strokes, and the letters seem to touch each other. Commercial Script and Zapf Chancery are examples.

ABCDEFGHIJKLMNOPQRSTUVW
abcdefghijklmnopqrstuvwxyzabcdefghij
OLDSTYLE   GARAMOND

ABCDEFGHIJKLMNOPQRSTUV
abcdefghijklmnopqrstuvwxyzabcdefg
MODERN   TIMES ROMAN

ABCDEFGHIJKLMNOPQRST
abcdefghijklmnopqrstuvwxyza
SQUARE SERIF   CLARENDON

ABCDEFGHIJKLMNOPQRSTUVWX
abcdefghijklmnopqrstuvwxyzabcde
SANS SERIF   HELVETICA

ABCDEFGHIJKLMNOPQRS
abcdefghijklmnopqrstuvwxyzabcdefg
SCRIPT   COMMERCIAL SCRIPT

𝕬𝕭𝕮𝕯𝕰𝕱𝕲𝕳𝕴𝕵𝕶𝕷𝕸𝕹𝕺𝕻𝕼𝕽𝕾𝕿𝖀
abcdefghijklmnopqrstuvwxyzabcdefg
TEXT LETTERS   OLD ENGLISH

ABCDEFGHIJKLMNOPQR
abcdefghijklmnopqrstuvw
DECORATIVE   COMSTOCK

Text letters resemble the hand-drawn letters of the early scribes. It is usually selected for religious documents, certificates, diplomas and invitations, and is rarely used otherwise. Old English and Engravers' Text are examples.

Decorative types are novelty styles or faces and used primarily to command attention. They are generally contemporary faces and do not fit any of the standard classifications. Designed to express different moods, they may be eccentric in appearance. Comstock is one example.

Type Families

Some types have many variations, and these various styles are said to be in the same family. Examples of these variations in a type style are: *light face, medium, bold, extra bold, expanded* and *condensed,* with italic versions of each in most cases.

A FAMILY OF TYPE

Helvetica Light

Helvetica Light Italic

Helvetica Regular

Helvetica Regular Italic

Helvetica Medium

Helvetica Medium Italic

Helvetica Bold

Helvetica Bold Italic

Helvetica Black

Helvetica Black Italic

Helvetica Regular Condensed

Helvetica Regular Condensed Italic

READABILITY AND LEGIBILITY

Most people who are not concerned with the fine points of typography use readability and legibility synonymously. But there is a distinct difference: readability is the ease of reading a printed page, whereas legibility refers to the speed with which each letter or word can be recognized. Readability refers to the type arrangement; legibility is concerned with type design.

Readability and legibility are dependent upon several factors that must be considered when selecting a typeface. These include texture and finish of paper, color of ink, typeface, size of type, line length, line spacing, etc. Type should be set to be read with little effort or eyestrain. Proper line spacing is important to the appearance of an ad or a printed piece. Each job presents a different problem, depending on the type style used, whether capital letters or lower-case letters are to be used, etc. Good typographers will give that little extra service, in the best interests of their customers, to be sure that the spacing enhances the typography. In practice, good spacing is often a matter of common sense.

NO LETTERSPACING

LETTERSPACING IS THE AMOUNT OF

2 POINT LETTERSPACING

LETTERSPACING IS THE AM

4 POINT LETTERSPACING

LETTERSPACING IS THE

Letterspacing is the amount of space used between letters, negative or positive, either for readability, aesthetics or to fill a certain area. It is used mostly in capital letters for display with small amounts of positive letterspace, or in "tight" typography or *kerning,* with negative letterspace. Some lines require corrective letterspacing to make all letters appear optically evenly spaced.

Negative letterspacing involves the removal of space between letters individually (kerning) or between all letters equally (called *white space reduction* or *tracking*). The most common kerning combinations are we, We, yo, Yo, ve, Vo, wa, Wa, Ta, To, ye, Ye, wo, Wo, va, Va, WA and VA.

Line spacing is the amount of space between lines which is known as leading (pronounced *ledding*) and is always expressed in points or fractions of a point. There is no set rule to follow. Too much leading can sometimes be as bad as not enough. Typefaces with long ascenders and descenders require more leading. Also, the wider the measure of text composition, the more leading is required for good readability. Leading is measured from baseline to baseline.

NO LINE SPACING (ALSO CALLED "SOLID")

The amount of space between lines is known as leading. There is no set rule to follow. Too much leading can sometimes be as bad as not enough. Typefaces with large

1 POINT LINE SPACING

The amount of space between lines is known as leading. There is no set rule to follow. Too much leading can sometimes be as bad as not enough. Typefaces with large

2 POINT LINE SPACING

The amount of space between lines is known as leading. There is no set rule to follow. Too much leading can sometimes be as bad as not enough. Typefaces with large

PRINTERS' MEASUREMENTS

The *point* and the *pica* are two units of measure universally used in printing in most English-speaking countries. Their use is primarily in typesetting. Type size is measured in points. Line length measure is in picas and points. The pica is used to express overall width or depth as well as the length of a line.

The point measures .0138 or approximately 1/72 of an inch. In other words, there are 72 points to the inch. All type is designated in points (10-point Caslon, 24-point Baskerville, etc.). Points are always used to specify the *size* of type. Typefaces may be set in sizes from four to 144 points, but are generally used in six to 72 points. Line spacing is also specified in points (two points of leading, etc.).

The pica is used for linear measurements of type. (A pica gauge is the printer's measuring tool.) There are 12 points to 1 pica, and approximately 6 picas to 1 inch. The length of a line is specified in picas, as well as the depth of a type area. For example, a given block of copy is to be set 20 picas wide by 36 picas deep. Inches are never used in type measurement.

The em is also important in typesetting although not a part of the point system. It is the square of the type size (a 10-pt. em is 10 points wide and 10 points high) and is used for measuring the quantity of type. The most common use of the em space today is that of paragraph indention. An *en* space is one half of an *em;* a *thin* is either ¼ or ⅓ of an em space. These fixed spaces are used for tabular composition.

The agate line is a measurement used by newspapers to sell advertising space. There are 14 agate lines to an inch. An agate line refers to the space occupied by one line of agate type in one column. The width of the column can vary from paper to paper. A 60-line ad can take several forms: 60 agate lines in one column, 30 agate lines in 2 columns, etc.

PROOFREADERS' MARKS

The proofreaders' marks shown on the following pages are standard and should be familiar to everyone working with type. It is important to use these accepted signs, rather than others which will not be understood by the typesetter. Marking changes with a colored pen or pencil enables the typesetter to see the corrections more easily. The illustration on page 45 shows how these symbols are used in actual practice.

Delete and Insert

ℓ   Delete, take out

℮   Delete and closeup

ℓ/ℓ   LETTER SP A CE

Insert space(more space)

☐   Em quad space or indention

stet   Let it stand—(all ~~matter~~ above dots)

Punctuation Marks

⊙   Period

∧   Comma

⦿   Semicolon

⦿   Colon

𝒱   Apostrophe or 'single quote'

𝒱   Open quotes

𝒱   Close quotes

?/   Question mark

!/   Exclamation point

=   Hyphen

(/)   Parentheses

/−/   Dash

Style of Type

wf   Wrong style of type

lc   Set in LOWER CASE or LOWER CASE

caps   SET IN capital letters

caps+lc   Lower case with Initial Caps

sc   SET IN small capitals

rom   Set in roman type

ital   Set in italic type

lf   Set in (**light face**) type

bf   **Set in** bold face **type**

Spacing

⌒   Close up entirely; take out space

‿   Less space between words

Insert space

Paragraphing and Position

⌐   Move to right ⌐

⌐   ⌐ Move to left

⊔   Lower (letters or words)

⊓   Raise (letters or words)

=   Align horizontally

‖   Align vertically

¶   Begin a paragraph

no ¶   No paragraph.

run in   Run in

flush ¶   No paragraph indention

tr   Transpose letters in a word

tr   Transpose enclosed in circle (matter)

Miscellaneous

✕   Broken type

⁹   Invert (upside-down type)

↓   Push down space

(*sp*)   Spell out (Capt.) Smith)

ok w/c   OK "with corrections"

ok a/c   or "as corrected"

⟨ ⋯ ⟩   Ellipsis

Proof with Errors Marked

cap THE PRACTICE OF TYPOGRAPHY, if it *be* followed *tr*
faithfully, is hard work—full of detail, full petty *of*
restrictions full of drudgery, and not *Greatly* *lc*
a rewørded as men now count rewards: There are ⊙
times when we need to bring to it, all the history
and art and feeling that w can, to make it beara- *e*
ble.

no ¶ But in the light of history and of art, and of ⋀
knowledge and of mans achievement, it is as
interesting a work as exists—a broad and
humanizingemployment which can *indeed* be *rom*
eq # followed merely as a trade, but which if per-
wf fected into an art, or even broadened into a pro- ⌐
fession, will perpetually open new horizons to
tr eyes our and opportunities to our hands.

—D. B. Updike *sc*

Proof after Corrections Have Been Made

THE PRACTICE OF TYPOGRAPHY, if it be followed
faithfully, is hard work—full of detail, full of petty restrictions, full of drudgery, and not greatly
rewarded as men now count rewards. There are
times when we need to bring to it all the history
and art and feeling that we can, to make it
bearable. But in the light of history, and of art,
and of knowledge and of man's achievement, it is
as interesting a work as exists—a broad and
humanizing employment which can indeed be
followed merely as a trade, but which if perfected
into an art, or even broadened into a profession,
will perpetually open new horizons to our eyes
and opportunities to our hands.   —D. B. UPDIKE

TYPESETTING

Before digital typesetting there were three basic methods of producing type: (1) cast metal or hot type composition which refers to cast metal type set by hand or machine, (2) typewriter or strike-on composition (sometimes called *cold type*) and (3) photographic and electronic typesetting.

Hand-set composition is produced with individual metal characters assembled into lines using a composing stick much as Gutenberg did in 1450. The composing stick is held in one hand while the letters are selected from a type case and placed in the stick until a full line is set.

Machine composition Machine-set copy can be produced on any of several machines. The most used were Linotype and Intertype machines which cast a line of type at a time, and the Monotype system which was a combination of two machines, a keyboard and type-caster that cast one character at a time. While these machines are still used in some parts of the world, they have been almost completely replaced by digital typesetting in North America, Europe, Australia, Japan and South Africa.

Typewriter or cold type composition As offset lithography was growing during the 1950s and 1960s, alternative methods for typesetting were developed. One of these was *strike-on* or *typewriter composition*. Typewriter methods were a forerunner of word processing systems that were introduced during the 1970s.

Photographic and electronic typesetting Three distinct generations of photographic electronic typesetting evolved. All were characterized by the need for master character images with systems for character selection, exposure and positioning.

(1) First generation devices were based on the linecaster mechanism. Speed was very slow at about five 30-character lines per minute.

(2) Second generation phototypesetters were electromechanical. These machines essentially replaced hot metal during the 1960s and 1970s. Average speed was about twenty 30-character lines per minute.

(3) Third generation phototypesetters applied cathode ray tubes (CRTs) which selected characters from memory and drew them on the CRT. These devices could operate at speeds of 500 to 1,000 30-character lines per minute.

DIGITAL TYPESETTING

Digital Fundamentals

Since the early 1980s, printing and publishing technology has been evolving digital methods for production. The *bit* becomes the building block of digital imaging. A "1" indicates it is on; a "0" indicates it is off. Bits are organized as *bytes* (8 bits to a byte is standard but new systems use 32-bit bytes) which contain significantly more information. In digital computing, bits and bytes form the underlying infrastructure for memory, data storage and actual computing.

The same on or off concept is the foundation for today's graphic arts industry. Scanners analyze images *spot* by *spot* to record the absence or presence of data; digital laser and inkjet printers, imagesetters and platesetters print out pages by assembling spots to produce type, graphics and pictures. Most digital imaging devices use a laser. The laser creates a *spot* whose size is based on the resolution of the device. This is the basic laser *spot* and its diameter is measured in microns— hundred thousandths of an inch.

In the printing world pictures are reproduced on printing presses as patterns of *halftone dots*. These are clusters of spots that fool the eye into perceiving a level of gray. Thus a *dot* should refer to a halftone dot. But many use the terms dot and spot as though they were the same. It takes a group, 10 or more, of the laser *spots* to make a halftone *dot*.

Spots are on or off, black or not there. *Pixels* are generated by video monitors or some specialized recorders, and they are *spots* with varying levels of energy, thus creating different shades of gray. A pixel can be 100% dark or at some percentage of gray. Thus red, green and blue pixels can be combined at varying levels to display a color picture. The term *pixel* is often misused for *spot* or *dot*.

Digital Resolution

An important factor concerns the reproduction of photographs and the relationship between spots per inch and halftone dots, or line screen. To create a halftone dot, that is, a specific level of gray (at least 100 gray levels) as perceived by the human eye, a matrix of about 10 x 10 of the printout unit's spots is used. This 10-to-1 ratio provides a rule of thumb for determining the possible line screen that a particular printer might produce. A 300 spi printer can only output a 30-line screen with 100 gray levels; a 1,000 spi printer can output a 100-line screen. Thus, to produce 133- to 200-line screens, which are typical of the graphic arts, one needs printers with 1,300 spi or greater.

Lower resolution printers can simulate pictures by creating artificial gray levels through a process called *dithering*. But as the link between electronic and desktop publishing and the graphic arts world increases, the need for higher resolution output is apparent.

Rasterization and Imagesetters

The Monotype Lasercomp in 1978 and the Linotype Linotronic in 1980 used lasers to output grids or raster images of type on film for printing. These devices output complete pages of text and images. Because they could produce images as well as type, they were called *imagesetters* rather than typesetters. Although there are still 2nd and 3rd generation phototypesetters in use, most have been replaced by imagesetting technology.

Digital Type

All output devices today are raster-based. This means that they create type and images as patterns of spots or dots on paper, film, plate and other substrates.

In 1985, Adobe introduced PostScript® as a language for driving raster-based output devices, an interpreter for driving raster-based output devices, and for producing typefaces as vector-based outlines. Today, over 100,000 fonts are in PostScript form. All digital fonts now fall into four categories.

- *PostScript* or *Type 1* fonts are scalable outline fonts which are defined using PostScript's bezier curves and work best with Raster Image Processors (RIP) because they do not need to be converted to be ripped and output.

- *TrueType* fonts are also scalable outline fonts but they are based on quadratic curves, not bezier curves. Created by Apple and Microsoft, these fonts must either be converted to Type 1 before being ripped or a TrueType Rasterizer must be used to create the bitmap for the output device. This conversion to Type 1 or rasterizing is often invisible to the user.

- *Open Type* is a new font format created by software producers to establish a standard outline for type used for print or electronic publishing.

- *Bitmap* fonts are non-scalable pixel maps of a given type face. To use bitmap fonts, a separate font file must be used for each size, or "jaggies" will appear on type.

Media for Digital Information Storage

Tape In reels, cassettes or cartridges. High density but *serial* in nature (one code after another) requiring winding and rewinding.

Floppy disk 5¼" and 3½" formats, from 360k to 2 mb. Random access and easily transportable.

Rigid (or hard) disk 8", 5¼" and 3½" formats, from 360k to 2 mb, from megabytes to many gigabytes. Various formats of rigid disks are based on manufacturer formats such as:

Magneto Optical (or MO): 128 to 256 mb.
Zip: 100 mb.
Jaz: 1 gb.

CD-ROM Compact Disk–Read Only Memory. Optical in nature. Capable of 600+ megabytes of data (millions of bytes).

DVD *Digital Video* (or Versatile) Disks (DVD) are record-only plastic platters like CD-ROMs with capacities starting at 4 gb and expected to go over 10 gb. They can record and playback audio, video and computer information. They are expected to replace CD-ROMs.

Data (or file) management allows jobs to be referenced by name or number, and the system will store them, later retrieving them by the name or number requested.

With the addition of memory management, the early cluster systems were able to incorporate more sophisticated *programs* (or *software*) for typesetting and composition. These programs performed hyphenation and justification, pagination and other functions, and thus *computer typesetting, multi-terminal, data management* systems were born. Since it was the "front end" of the typographic printout process, the term is often used.

Disk array A disk array connects two or more disks through a single controller. To the computer, the array looks like a single volume. The controller shuttles data between each disk. While one disk is writing a block of data, the other disk is available for the next block. In this way, data transfers can be two or three times faster than with a single disk.

Disk performance is directly related to how fast the disk spins (rotation rate) and how quickly the drive head can seek, access, and transfer data.

- Access time is the amount of time it takes to drive data head to start reading/writing data from within the track.
- Seek time is the amount of time it takes to drive data head to get to the requested track.
- Burst speed is the fastest speed at which the drive can read from or write to the disk.
- Sustained transfer rate is the maximum rate of data transfer a disk can maintain, as the data heads move across the disk.

Typesetting Commands

Batch text input consists of two kinds of information: the *characters* you will see upon output, and the *commands* that determine how you will see them. Commands include changes in point size, typeface, line length, leading (line spacing), positioning, tabular, columnar makeup, indention and others. Commands may be *interactive* or *code* oriented.

Code commands may be preceded by a single key. This is called a *fixed field* approach, since the letters and numbers must be exactly right (for instance, point size would need two digits, thus 9 point would have to be 09, etc.). Variable field commands have a beginning and ending key, such as the open and close brackets, thus allowing more variation in the data within.

TYPICAL COMMANDS

*il	Indent Left	*i	Italic
*ir	Indent Right	*cc	Change Column (measure)
*ib	Indent Both (sides)	*cf	Change Font
*rr	Ragged Right	*cl	Change Leading
*rl	Ragged Left	*ts	Tab Set
*cp	Change Point (size)	*ep	End Paragraph

Commands can be almost unintelligible if the letters used are not memory oriented *(mnemonic).* Too often, the programmer tried to stay with single letters, and then ran out. Newer systems start with multiple letters or multiple precedence keys and are more user friendly — although CMD/CATL + OPT/ALT + Shift is no fun.

A present approach to commands for typesetting is called *generic coding.* Typed information is identified as *[text]* or *[head]* and then translated into the appropriate typographic parameters. Thus, authors and other editorial originators can code data by its generic appearance without regard to specific typographic specifications. Standard Generalized Markup Language (SGML) can generically encode text as well as define "what" the text is, such as defining a number as an amendment to a bill, which would allow searches only on amendments.

Interactive commands are based on the pop-up, pull down, or selectable *menus* which list the various typographic and functional alternatives available. The user defines the text or graphics on the screen by pointing at it while pressing a special key or button. That material is then highlighted as a reverse, tint or color. The proper menu is chosen, and the individual function within it is selected. Usually there are two approaches: a simple button for selection or a *dialog box* for the insertion of specific

A TYPICAL MENU WITH SELECTABLE FUNCTIONS

```
┌─────────────────────────────────────────────────────┐
│           Hyphenation & Justification                │
│ ┌─Auto Hyphenation──────┐ ┌─Justification Expansion Method─┐
│ │ ◉ Off  ○ On           │ │ ◉ Standard      ○ Uniform      │
│ │ Smallest Word:  [6   ] │ │ ○ Spaces only   ○ Other:       │
│ │ Break After:    [3  ]  │ │        Spaces: [16 ]           │
│ │ ☐ Break capitalized words│ │       Overall: [7  ]          │
│ └───────────────────────┘ └─────────────────────────────┘
│ Hyphens in a Row: [unlimited]    ( OK )    ( Cancel )       │
└─────────────────────────────────────────────────────┘
```

values or data. The highlighted material is then affected, and the result is immediately displayed on the screen.

It is common to see both menu and code approaches in use within the same system or program. Many menu selections have key equivalents. For instance, the change of highlighted copy to bold type could be accomplished via the menu or by keying **[Command Key] b**.

Almost all systems include the ability to see type and graphics on screen in simulated page form with all items in position. This can be either in a preview mode, where coded text is translated into a screen image, or interactive, where one deals in a typographic context at all times. The interactive integration of text and graphics has no single descriptive name; hence, the term *publishing* is often used as a synonym.

Word Processing (WP)

In business offices, word processors began as typewriters connected to some form of recorded medium. This permitted typists to *capture* keystrokes for editing and correction. The IBM MT/ST was the first word processor and was essentially a typewriter with magnetic tape cartridges that allowed the operator to update and change input copy during or after typing.

Most word processors are personal computer-based and can perform formatting and typographic page layout. Most important, the text being handled may eventually find its way to typesetting. Having the recorded medium thus allows input to typesetting without re-keyboarding.

Output from word processors is usually typewriter-like. Newer high-speed printers using impact, ink-jet or electrostatic techniques can simulate typesetting. The electronic nature of word processors links nicely with electronic reproduction systems.

Word processors evolved into *stand-alone* (self-contained) units or networked systems, but are now replaced by programs in personal computers.

Personal Computers

Usually consisting of a keyboard, a video screen and a floppy disk (and internal microprocessor computer), personal computers, also called workstations, have become the Model Ts of the electronics world. Easy-to-change software programs allow each unit to perform word processing, accounting and other functions for business or personal use. Almost all PCs are *multitasking* in that they can perform more than one function at a time. Thus, while a file is printing, the user can be editing or performing other functions at the same time.

Telecommunications

Information stored in most electronic text systems may be communicated over telephone lines. *Modems* convert signals from these systems into telephone signals and then re-convert them at the other end. Transmission requirements are called *protocols,* allowing transmission in one direction *(asynchronous)* or in both directions simultaneously *(synchronous)* for error correction. All word processors and personal computers telecommunicate.

Interfacing

The linking of different devices is called *interfacing.* It may be accomplished by reading a recorded medium, accepting telecommunications, or simply connecting the units by cable. A *converter* is required to translate the *coding system* of one system to another. A *translation table* establishes individual conversions of input codes into new output codes.

Desktop Publishing

The multi-purpose nature of the personal computer, the proliferation of the typographically-oriented plain paper laser printer and the increasing pagination ability of word processing programs led to the application called *desktop publishing.* The essential part of this approach is the use of typographic instead of typewriter characters and the integration of boxes, shapes, line art, low-resolution pictures and other graphic elements into page form. Today desktop publishing is common for all aspects of prepress, print and electronic publishing production.

It is more appropriate to define this application as *page processing,* since its approach to putting pages and documents (combinations of pages) together is applied at all levels of workstation and system configurations. The use of high-level graphics-oriented workstations with *multi-tasking* functions permits increased productivity. UNIX is a well-known multi-tasking operating system, and other PC-based operating systems have evolved to multi-tasking.

Mnemonics

Mnemonics means memory oriented. For example, you can remember the colors of the spectrum with the name ROY G BIV for Red, Orange, Yellow, Green, Blue, Indigo, Violet. Many typesetting commands are mnemonic in nature.

Most desktop programs have keyboard shortcuts or sequences for program functions. Almost all use a combination of Command or Control, Alt or Option, or Shift keys to select or activate the function. They are followed by a single letter, which may reference the function in a memory-oriented manner: B for Border but Shift B for Bold, C for Copy but X for Cut. With single letter mnemonics there are only 26 chances for logic. After that the shortcuts may not be so short.

Networking

Workstations may be linked by special cable so that peripheral devices may be shared. Local Area Networks (LANs) allow for high-speed movement of text and graphics data. A central printer is then called a *print server* and a central magnetic disk for storage is called a *file server.* Today's systems are combinations of intelligent workstations connected on networks with a variety of peripherals.

The Internet is a network of networks that links workstations over telecommunications lines to share files and exchange e-mail internationally. The World Wide Web is a part of the Internet that lets users present information formatted with typographic text, color and images on desktop computer monitors using *browser* software, and then *hyperlink* between documents, files and user sites by clicking on designated works or images.

HTML

Instead of using specific typographic commands, some document creation systems use generic tags to define the format of typographic pages. SGML is a standardized approach to defining the *look* or format of a document. A derivative of SGML is HTML (Hyper Text Markup Language), the document markup language for the formatting of information on the World Wide Web.

XML

The Extensible Markup Language (XML) is a subset of SGML that defines a class of data objects and describes the behavior of programs that process these objects. XML extends SGML to the Web with interoperability with HTML.

Printers are using the World Wide Web to promote themselves, accept files from customers for printing, provide job tracking information and quote on job specifications. Many desktop application programs allow HTML conversion as a derivative of files created for printing.

HyperText Text and graphics in Web documents can be linked to text and graphics in other Web documents, and you can jump between them with a mouse-click. HyperText links are displayed in a color other than the standard body text.

HTML looks like a print-based markup language. HTML lets you tag page elements for specialized formatting: headers, list items, italics and so forth. You may also embed links to other documents, graphics, web servers and so on. Here is an example of HTML coding:

```
<HTML>
<HEAD><TITLE>Rochester Institute of
Technology</TITLE></HEAD>
<BODYbgcolor="#c6c6c6"
background="Images/Backgrounds/greydiagRIT.gif">
<p align=center>
<A href="/Imaps/RITHome.map"> <img ISMAP
alt=width=466=height=252
border=0 src="/Images/Buttons/RITHome.gif"> </A>
<p><table width=100%><TR><TD align=center>
<A HREF="/Phone/"><IMG ALT="People" border=0
SRC="/Images/Buttons/mainbar_people.gif"></A></TD>
```

More information relating to Digital Imaging, Desktop Publishing, Imagesetters, Electronic Editing, Networking and Digital Typesetting is in the section on *Digital Prepress (page 91).*

COPY AND ART PREPARATION

Just as digital imaging has changed the mechanics of typesetting, copy and art preparation have been affected similarly. At one time, manuscripts were created on typewriters which then needed to be converted to type prior to typesetting and reproduction. This required *markup* of the copy to indicate typographic style. Markup is still required to indicate typographic and editing instructions on copy from any output device for proofreading and correction.

TYPE SPECIFICATIONS AND COPY MARKUP

The type specifications should be clearly and completely written on the typed manuscript (not the layout) with the size, leading, typeface and measure in that order. In marking the size, the leading (space between lines) should always be specified in the form of a fraction — 8/9 (8 on 9) means 8-point type with 1-point leading, 10/10 means 10-point solid (no leading), etc. Leading does not affect the number of lines, only the depth. When a large amount of copy is to be set, care should be used to select a typeface that does not require much leading.

Always specify line measure in picas. An example of complete specifications would be: 10/12 Times Roman x 20, which means 10-point Times Roman with 2-point leading to be set in a column 20 picas wide. Indicate whether it is to be justified or set flush with a ragged right or left margin.

Careful word spacing is important! It must be remembered that all formulas for copyfitting are based on even word spacing throughout. If a job is widely word-spaced, copyfitting calculations may be upset. Many typists type two word spaces at the end of a sentence. This should not be done as it affects word spacing calculations.

All paragraph indentions, hold-ins, etc., should be indicated in *ems* of the type size — not in picas. (An em is the square of the body type, i.e., an 8-point em is 8 points square.) Be sure to indicate whether paragraphs should be indented or kept flush. Mark headings flush left, flush right, centered and/or indented according to the design.

Underscoring a word or line has a very definite meaning to the typesetter. One underscore means *set in italic,* two mean SET IN ALL SMALL CAPS and three underscores mean SET IN ALL CAPS. Underscore also means "underline" which is done with a rule by typesetter or drawn on a repro, not to be confused with setting italic. Although a wavy underscore means **set in bold face,** it is always better to mark "bf" in the margin.

CONVENTIONAL ART PREPARATION PROCEDURES

Art and copy (not to be confused with manuscript copy) are terms used to describe *all* material supplied for reproduction. It includes not only the type, but also diagrams, drawings, photographs and color transparencies. Art preparation embraces all of the steps in getting the art and copy ready for reproduction. There are two basic steps: (1) the design or layout of a printed piece, and (2) the preparation and assembly of the various components for reproduction, either manually or electronically.

The Layout

The first step in designing a printed piece is preparing a layout, a blueprint of a printed job. It is important that the layout person know the purpose of the printed piece (as well as the printing process to be used) so that the layout will reflect this. The layout may be a very rough visual, a loose comprehensive or a tightly rendered comprehensive, looking like the finished job in all details. The final layout may be crudely drawn, but it must be accurate in size and accurately marked, as it is the blueprint from which all people (including the designer) producing the job will take their specifications.

At the same time the layouts are being prepared, a blank paper dummy of the job should be made to size, preferably on the stock to be used. This will help the designer visualize the final appearance, enabling him/her to provide proper margins, bleeds, color, etc. Attention should be paid to staying within the boundaries of standard paper sizes and the printing process.

With the approved layout and paper dummy as a guide, the artist is ready to put all the elements together into a final pasted-up *mechanical* (or *pasteup*) for graphic reproduction or in an electronic layout on a desktop workstation.

ROUGH LAYOUT

COMPREHENSIVE LAYOUT

LINE

HALFTONE

Kinds of Original Images

There are several kinds of original images. In general, they are classified according to whether the copy is *line* as in type matter, diagrams, and pen and ink drawings; or *continuous-tone* as in a photograph with a variety of tones. These are further broken down as to whether they are to be reproduced in one color, multicolor, or process color, and whether they are alone or in combination (line and continuous-tone).

Therefore, we can have either line or continuous-tone images in one color, combinations of line and tone in one color, line or tone in more than one color, combinations of line and tone in more than one color, process color tone images, and combinations of process color, tone and line.

Continuous-tone images may be either rendered illustrations or photographs. For reproduction by most printing processes,

HALFTONE DOTS ENLARGED

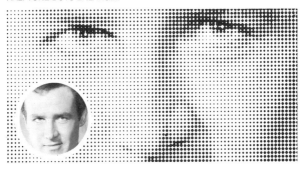

continuous-tone images are converted to dot pattern images, or *halftones*. Halftones have the appearance of continuous-tone images because of the limited resolving power of the human eye. This limitation accounts for an optical illusion. Small half-tone dots of screen ruling of 133 lines per inch (lpi) and higher, when viewed at the normal reading distance, cannot be resolved as individual dots, but blend into a continuous tone *(see illustration, page 57)*.

Preparing Art for Multicolor Printing

Art for multicolor reproduction by conventional printing is usually prepared in black and white on a mechanical. When hairline register is not required, the art for the key color is pasted to a sheet of illustration board, and the art for other colors are registered on clear acetate overlays hinged to the board. Color and screened percentages (if any) should be marked on each overlay.

For hairline register, all colors should appear on the same board, if possible, and the color break indicated on a tissue overlay. Pin register devices *(see page 74)* can be used to ensure hairline register of separate image elements if they cannot be combined on the same board. Finish size should also be indicated on the art. If several pieces of art are prepared for a job, they should all be drawn the same size. This simplifies camera operations and reduces preparation costs. All copy should be *keyed* or cross-referenced by page number, title or job number.

In preparing art for the camera, line and continuous-tone images should be separated. Full-color reproductions form a third group. Images not assembled together should be cross-referenced or keyed for easy identification. This can be done by making an outline drawing of the image on the mechanical or by photocopies which are pasted where key images will appear. Images keyed for color may be indicated by different color areas drawn on a tissue overlay on the mechanical.

Gravure requires extensive assembly of art and copy as each element is handled separately. Each method of cylinder preparation handles copy differently.

Screen printing art and copy can be prepared either manually with the knife-cut film method, photomechanically or by a combination of both these methods.

Register In preparing art where two or more colors are to be printed, register of the different color images is important. Jobs where color areas are independent of each other are considered *no-register* jobs. In 4-color process printing, *commercial register* must be precise (± ½ row of dots or less).

Special colors Additional colors and color values can be obtained by overprinting two or more inks. Overprinting can be in both solids and tints (various tones of solid color). Special colors like logos or corporate colors often cannot be reproduced by color overprints and must be printed with special color inks usually specified as *PMS colors.*

Duotones A black and white photograph can be reproduced in two colors to obtain more depth or contrast. This two-color half-tone is called a *duotone.* Two closely related colors, black and an accent color, or even two black inks can be used. The original picture is photographed twice, one negative emphasizing the highlights (lightest parts), and the other, the shadows (darkest parts). To avoid moiré patterns the darker image should be printed at 45° and the other color at 15° or 75°. Sometimes a two-color *duotone effect* is used by printing a screened tint of a color over a black halftone *(see illustration, page 82).*

Reflection and Transmission Art for color reproduction falls into two classes: *reflection* and *transmission* copy. Reflection copy is original material for reproduction which is viewed and photographed by reflected light, such as oil paintings and photographic color prints. Transmission copy, such as color transparencies and color negatives, are viewed and photographed by transmitted light. The terms reflection and transmission are also used to describe originals for scanning. The majority of copy for reproduction is in transmission mode — either positive color transparencies or negative color films.

Scaling and Cropping

Many different original art elements may be used in producing a printed piece. Some may have to be reduced or enlarged in size,

DIAGONAL LINE METHOD OF SCALING

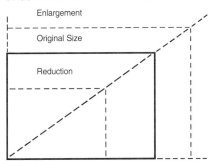

which requires *scaling* and *cropping.* Scaling has to do with changing the size of the reproduction without changing the ratio of the dimensions *(see illustration, page 59).* In addition to the diagonal line method of scaling, a printer's proportion scale can easily determine a reproduction size. Cropping is the term used to describe the process of trimming or cutting off parts of a picture or other copy element.

DIGITIZING COPY PREPARATION

Digital imaging makes it easier to use graphics, text and images so they are more common in the design of printed products. In fact most print designers use computers and software programs to compose pages, documents, promotional, presentational or utility printed products so that the digital design eliminates the need for most copy preparation. The copy preparation is essentially completed in the process of designing the product. The design is already in digital form so changes and/or corrections can be made digitally. In effect, the designer uses the computer and software to perform all the operations cameras are used for in conventional preparatory operations and most of the manual operations as well.

Almost all aspects of copy preparation are now performed with desktop publishing components such as personal computers using application software, desktop scanners and printers. There are software programs for *color retouching, color correction, trapping, page layout, page makeup, vector illustrations, drawing programs (vector oriented art), paint programs* and other image manipulation programs. Although conventional (mechanical) art preparation is still used, it is being replaced by totally electronic methods.

How this is accomplished is explained on pages 112-114 in the section on *Digital Prepress.*

IMAGING FOR GRAPHIC ARTS

Imaging is the most important step in the hierarchy of prepress operations. It is the operation in which images for reproduction are collected, created and/or corrected prior to assembly for platemaking. Since 1850, photography has been the principal imaging system for photomechanical platemaking. Two types of photography have been used: *creative photography* to produce original images for reproduction; and *graphic arts photography* to produce the films for making the plates for printing.

Since 1990, many photographic operations and supplies have been replaced by digital imaging systems. Digital cameras are being used for creative photography, and digital scanners, computers and software have been developed to produce digital files to make the films for making the plates, or for side-tracking the films and making the plates directly.

PRINCIPLES OF LIGHT AND COLOR

Knowledge of the principles of light and color is critical to the proper use of all visual processes including photography, printing and digital imaging systems. Light is the visible portion of the electromagnetic spectrum of radiant energy that varies in wavelength from the very short cosmic rays to long radio waves and electrical oscillations *(see page 84)*. The visible light spectrum varies in color from violet to blue, blue-green, green, yellow, orange and red. Wavelengths vary from 400 nanometers (nm) where violet borders on the invisible ultraviolet (UV) rays to 700 nm where red borders on the invisible infrared. (A nanometer is one ten-millionth of a centimeter.) *(See page 85.)*

Light has a number of important properties besides wavelength. Its rays can be transmitted, absorbed, reflected, refracted (bent) and polarized — all properties important to the appearance of objects and to photography and graphic arts. In addition, light can vary in intensity and color composition.

Color is a combination of the physical sensation of light and the psychological interpretation of it. Physically, the combination of all wavelengths of visible light from 400-700 nm with equal energy distribution produces the visual effect of white light. Also, physically, the color spectrum consists of almost 10 million perceptibly different colors.

Additive primaries Psychologically, the human eye divides the visible spectrum into three broad bands of color — red, green and blue, commonly abbreviated as RGB. The human eye contains three types of receptors, each sensitive to one of the primary colors of light. When the eye views a scene, the receptors are activated by the colors to which they are sensitive and they send impulses to the brain. The brain uses the impulses to re-create the scene. The three colors red, green and blue are called *additive primaries* because three lights of these colors when added together are seen as white light. These are the colors used for color separation filters *(see illustration, page 85).*

Subtractive primaries When an additive primary color is subtracted from white light, the two remaining colors form a new color. When red is subtracted from white light the new color is a combination of blue and green which is called *cyan* (C). When green is subtracted, red and blue remain, the combination of which is called *magenta* (M). When blue is subtracted, red and green light remain and the combination of the lights of these colors produces *yellow* light (Y). These are called *subtractive primary* colors which are abbreviated as *CMY*. The additive primaries (RGB) are the colors used in imaging, as the colors of the filters used for color separation, and for the phosphors in the video screens of computers, workstations and TV. The subtractive primaries (CMY) with black (K) added to become *CMYK,* are the colors of the process color printing inks on page 85.

PHOTOGRAPHIC IMAGING

Before the introduction of the scanner in 1950 all imaging functions for graphic arts were manual or by photography. Photography uses cameras and film, or paper bases, coated with a light sensitive emulsion of silver halide salts in gelatin or other binders. When the emulsion is exposed to light from a subject, scene or other image, the photons of radiant energy in the light reflected from or transmitted through the subject convert the silver halide salts in the emulsion contacted by light into a *latent image*.

When the emulsion is processed by special organic developers, the latent image is converted to varying amounts or densities of metallic silver corresponding to the amount of exposure received. The unexposed areas are converted to soluble silver complex salts in a fixing bath of sodium or ammonium thiosulphate (called *hypo*) that are removed in the wash water following fixing. A serious objection to the photographic process is the toxicity of the silver salts in the processing wastes which require special disposal usually by silver recovery systems.

Continuous tone vs. halftone Any image, photograph, wash drawing, pastel, oil painting, etc., consisting of a broad range of gradation of tones is known as *continuous tone*. In creative photography these different tones are represented by varying amounts of silver. The more silver, the darker the image, and vice versa. Letterpress and offset lithography are binary processes which can print only a solid density of ink in the image areas and no ink in the non-image areas. In order to reproduce the varying tones of a picture by these processes, graphic arts photography employs an optical illusion using a halftone screen. Halftone photography makes the reproduction of continuous-tone photographs possible by converting the image into a pattern of very small and clearly defined dots. The normal eye has a halftone screen resolution of about a 133 lines per inch (lpi). The dots in screens with lpi lower than 133, as in newspapers, can be resolved or seen by the normal eye *(see page 80)*.

Screenless printing processes are capable of printing varying ink densities to produce pictures having a wide range of tones without the need of halftones. *Gravure* uses cells with different shapes, sizes and depths, so different amounts of inks are transferred according to the tone values to be printed. Another process is *collotype* in which the image consists of reticulated gelatin that prints ink density in proportion to the amount of exposure the gelatin has received through a continuous-tone negative. *Screenless lithography* uses special plates made from continuous-tone films or FM halftones *(see pages 67 and 72)*.

Graphic arts cameras are being displaced by digital imaging. There are horizontal and vertical cameras. Horizontal cameras are built into darkroom walls. They consist of a *bed* with vibration-free suspension on which are mounted a *copyboard; bellows* with *lensboard* and high resolution, apochromatic (color corrected) *lens* with minimum aberrations and distortion; very high intensity *lights* such as *pulsed xenon* lights for color repro-

FLOOR-TYPE HORIZONTAL CAMERA

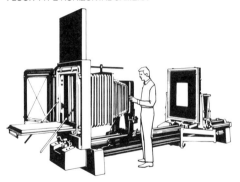

duction and *quartz iodine* lamps for single-color imaging; and *exposure* controls, usually *light integrators* to control exposures using analog computers to integrate the total quantity of light by adjusting exposure times as light intensity varies; all are in the camera area. The camera back with ground glass for focusing and vacuum back to mount the film for exposure is in the dark-room. Vertical cameras look like enlargers. They are smaller, save space and are usually in a darkroom.

Films, such as stable base polyester, are used where dimensional stability is critical, as in color reproduction. Special high contrast emulsions of silver halides in gelatin, known as lithfilm, are used for line and halftone photography. Continuous-tone film is used for color separations, masks and gravure printing. Special films are made for scanners, and new special daylight films are used to make contact prints for stripping or image assembly.

Films are color sensitized by using dyes in the emulsion. Ordinary or colorblind film is sensitive to ultraviolet (UV) and blue light. Orthochromatic film is sensitive to UV, blue and green light. Panchromatic film is sensitive to UV, blue, green and red light *(see illustration).* There are also films sensitive to infrared radiation.

COMPARISON OF COLOR SENSITIVITY
OF EYE AND PHOTOGRAPHIC FILMS

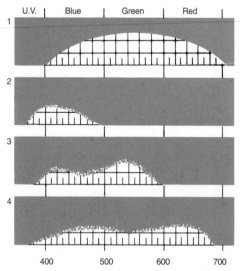

1. Sensitivity of the human eye
2. Ordinary color-blind film
3. Orthochromatic film
4. Panchromatic film

Development of latent images is done by immersing the film in a special developer. Each film type uses special developing agents and combinations of chemicals. Developing time and temperature are means of control used to get desired results. *Stabilization* processing used in typesetting eliminated washing by converting the unexposed silver halide to a stable complex which eventually stained on exposure to sufficient extraneous light. *Rapid access* processing using continuous-tone type developers is used to speed development of lithfilms exposed by laser-generated images. *Automatic film processors* are used to process all types of films. They not only save considerable time but produce more consistent results.

Dry films The pollution hazards of conventional silver halide processing has spurred the development of non-silver and dry processing photographic systems. A number of different technologies are used including: *chemical-free film* and a special thermal development; *dry multilayer film* that changes adhesion on exposure to a high power laser in a dedicated imagesetter; *thermal transfer* with a special device; *ablation technology;* and *dry silver* using thermal development in a special device. While all these systems eliminate the hazards of silver halide processing, most of them require dedicated imagesetters that are much more expensive than the film processors they replace.

Negatives and positives The usual product of the photographic process is a *negative* in which the light portions of the copy are represented by heavy or dark deposits of silver, and the dark portions of the copy are light or transparent. When negatives are printed on paper or film they produce *positives* in which the tone values are similar to the original copy.

Some platemaking processes require negatives, others require positives. Letterpress and flexographic plates are normally made from negatives. Positive films, and positive or negative prints are used for gravure. Some lithographic plate processes use negatives, others use positives. Screen printing requires positives. Negatives or positives can be line, continuous tone or halftone, and right or reverse reading on the emulsion side depending on the process and use. The use of computer-to technologies will eliminate the need for negative or positive films for platemaking.

Line photography consists of photographing solids, lines, figures and text matter. The copy is placed on the copyboard and focused on the ground glass in the back of the camera. The film is placed on the vacuum back; the lens aperture is set; the copy is illuminated by high-intensity lights; and an exposure is made through the shutter operated automatically by a timer or light integrator. Processing the film produces a negative.

Contact prints are used extensively in film assembly. They are made by placing a negative or positive over an unexposed piece of film in a vacuum frame, and exposing it to a special collimated point light source. The resulting contact print will be a positive if made from a negative, or a negative if made from a positive. It is possible to make negatives from negatives or positives from positives using special duplicating film. Most contacting is done with special daylight films with low sensitivity that allow contacts to be made outside the darkroom.

Halftone photography uses a halftone screen or electronic dot generation. The original glass halftone screen consisted of a grid of inked lines on two sheets of optically flat glass cemented together at right angles. The lines were approximately the same width as the spaces. Screens with rulings of 60- to 100-lines per inch (lpi) are used to make halftones for newspapers. Screens with 120- to 150-lpi are used for magazines and commercial printing, and 150- to 600-line screens are used for high-quality process color printing. The glass screen is practically obsolete, and all halftones now are produced by contact screens or electronic dot generation.

Contact screens are on film bases and are usually made from glass screens. Dots are vignetted with gradient density across each dot. Density is greatest at the center and least at the perimeter. There are gray screens with dots consisting of silver images. Dyed screens, usually magenta, contain dots in which the silver has been replaced by a dye. There are also square and elliptical dot screens for special effects, especially in the middle tones, and double-dot and triple-dot screens for special effects in the highlights *(see respi screen, terms section, page 229)*.

MAKING A HALFTONE NEGATIVE

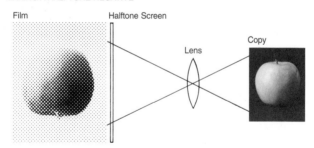

Film · Halftone Screen · Lens · Copy

Contact screen photography uses vacuum contact between the screen and the film. The variable density of the vignetted dots records variations as larger or smaller dots on the film, depending upon the amount of light reflected or transmitted from the copy. Contrast of reproduction can be varied by techniques known as *flashing* and *no-screen exposure*. Flash exposures are used to reduce contrast especially in the shadows by producing a dot over the whole film and are made by exposing the film to a yellow light. The no-screen or *bump* exposure is used for increasing the contrast in the highlights by removing the screen during a short part of the exposure. Additional control of contrast may be achieved with dyed screens by using colored filters during part of the exposure. Halftones made with contact screens have dots of different sizes with equal spacing between dot centers (AM screening).

Electronic dot generation (EDG) is done by special computer programs, called screening algorithms, for electronic scanners and imagesetters to produce halftone images. They combine, or assemble the pixels of digitized images into dots that simulate the dot sizes, shapes, screen rulings and screen angles of contact screen images which is called *amplitude modulated (AM) screening*. *Frequency modulated (FM)* or *stochastic screening* algorithms are also used by EDG or digital dot generators. They produce almost continuous tone images with very small dots of equal size and variable spacing. These images are independent of screen rulings and screen angles and are thus free of moiré patterns. AM and FM screening are discussed in more detail on pages 71 and 72.

AM SCREENING   FM SCREENING

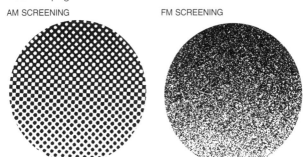

Contrast and Tone Reproduction

Using a stepped gray scale to represent the tones in a subject, normal tone reproduction results when the darkest area prints on the press as a solid and the lightest area prints without a dot.

The films may have very small unprintable dots in these areas which close up in the shadows and disappear in the highlights during printing. In AM screening intermediate tones have varying sizes of dots ranging from about 3 percent in the highlight areas to about 95 percent in the shadows. In FM screening all dots are the same size and the distance between them varies.

High contrast results when two or three steps in the shadow end print solid and several steps in the highlight end are white with a corresponding increase in density difference between other steps of the scale.

Low contrast results in AM halftones when the scale contains 80-90 percent dots in the shadows and 10-20 percent dots in the highlights with corresponding decrease in density difference between other steps in the scale. FM halftones have dot spacing corresponding to these tone values. A number of special techniques may be used to change the contrast in local areas of the reproduction.

COLOR REPRODUCTION

Color reproduction is based on the theory of three-color vision. It consists of *color separation, color correction,* and *color proofing,* and all the operations involved in carrying out these functions.

Color Separation

The process of color separation is analogous to seeing by the eye *(see page 61)*. The original photo or artwork is photographed using white light and three filters, each corresponding in color and light transmission to one of the additive primaries (RGB).

Placing a red filter over the lens produces a negative recording of all the red light reflected or transmitted from the subject. This is known as the red separation negative. When a positive or print is made from this negative, the silver in the image areas corresponds to areas that did not reflect red, which are blue and green. In effect, the negative has subtracted the red light from the scene and the positive is a recording of the blue and green in the scene which is called *cyan.* The positive is the *cyan printer.*

Photography through the green filter produces a negative recording of the green in the original. The positive is a recording of the combination of the other additive primaries, red and blue, which is called *magenta.* The positive is the *magenta printer.*

The blue filter produces a negative recording of all the blue in the subject. The positive records the red and green which when combined as additive colors produce *yellow.* This positive is the *yellow printer.*

These colors, cyan, magenta and yellow (CMY), are called *subtractive primaries* because each represents the two additive primaries left after the third one has been subtracted from white light. These are the colors of the inks used for process color reproduction. These colors are illustrated on page 85.

When the three positives are combined in printing, the result should be a reasonable reproduction of the original. Unfortunately, it is not. The colors, other than yellows and reds, are dirty and muddied. Usually there is too much yellow in the magentas, purples and blues, and too much magenta in the greens, blues and cyans. This is not a flaw in the theory but is due to deficiencies in the colors of the pigments used in the inks.

Corrections must be made in the color separations because of limitations in the colors of the inks. Even after these corrections are made, the printed result is not satisfactory. It lacks saturation and contrast. A fourth, *black printer,* is added to overcome this. It improves the contrast of the grays and deep shadows and increases their saturation. It may be a *skeleton* or a *full* black. There is a trend toward the use of full blacks. Other colors are reduced proportionally so that inks transfer or *trap* properly on high-speed presses. This operation of reducing colors and printing a full black in shadow areas is called *undercolor removal* (UCR) and is used extensively on multicolor presses.

Gray component replacement (GCR) is like undercolor removal (UCR). Wherever three color inks overprint, the two predominant ones determine the hue of the color and the lesser third color indicates the final color's grayness or saturation. An amount of the three colors equivalent to the lesser third color (called the *gray component*) can be removed from the color and replaced with black ink. Like UCR, different amounts of GCR can be used. The amount of color added in the shadows to increase their saturation is called *undercolor addition* (UCA). Not only does this new technique save on ink costs, it increases color saturation, speeds makereadies on the press, and improves print consistency.

Color Correction

Color corrections to compensate for the spectral errors in inks and exposure errors in photography are made manually by *dot etching*, photographically by *masking* and electronically by *scanning*, and software.

Dot etching corrections are used to increase or reduce color in local areas of halftones using chemical reducers manually. Reducing dot sizes in negatives increases color and in positives reduces it. *Dry dot etching* color correction changes dot values by exposure and development using computerized exposure and development.

Masking is photographic color correction during color separation using low density negatives, made with special filters, in contact with color transparency originals or on a camera vacuum back with other originals. Practically all color separations are made on electronic scanners. Very few separations (less than 5%) that are made photographically use a direct screening method in which the original is masked and screening is done in a camera, enlarger, or by contact in a vacuum frame.

Electronic Scanning

This is used almost exclusively for image capture to produce color separations for color reproduction. Two configurations are used: (1) *drum* scanners using RGB filters and *photo multiplier tubes* (PMT) as sensors; and (2) *flatbed* scanners using RGB filters and *charge coupled devices* (CCD) as sensors. Drum scanners are more expensive but have been preferred because they had higher *resolution* — (8-12 bits per pixel) — and *dynamic range* from 3.5-4.0. CCDs originally had lower *resolution* and *dynamic range* of about 2.0 but newer developments have increased resolution to 8 bits per pixel and dynamic range to over 3.0 *(see page 93)*. On drum scanners the copy is mounted on transparent drums which can be used for transparencies or reflection copy. Flatbed scanners are either reflection or transparency types. A few can be used for both.

Analog scanners Original scanners were device dependent *analog* systems using RGB light beams through *PMTs* to record light intensities from the copy and send the beams to four computers that corrected and converted the RGB beams to *CMY* and the fourth computed the black separation from the RGB signals. After all corrections and computations were made, four beams representing CMYK images were sent direct to built-in film recorders that exposed films of the four color separations.

Digital scanners are device independent systems that record in the RGB color space and software programs are used to convert the RGB color space to CMYK separations that are used for printing. Calibration systems are needed to assure consistent conversion from RGB to CMYK color spaces. Digital scanners process RGB signals and store them in digital formats which can be used in a number of outputs. The most popular file is *PostScript* and outputs can be *printers* to view the file, *imagesetters* to produce films for proofing and assembling them into signature form for platemaking, *imposetters* for making fully imposed films for platemaking, *platesetters* for making plates directly without the use of films, and a file for producing images in *multi-media* like the *world wide web*. All these devices are described in the *Digital Prepress* section, page 91.

Digital cameras record images as electronic voltages that are converted to digital signals which can be used to replace the need for scanning. **Photo CD** can also replace the need for scanning. It is a system for converting conventional color transparencies and color negatives to *Image Pacs* that can be stored on CD-ROM discs which can be read on CD-ROM drives. The images in a special *YCC* color space can be reproduced as images in the *RGB* or *CMYK* color spaces for use in imaging software programs like regular digital separations without the need of a scanner.

AM/FM HALFTONES, SCREEN ANGLES AND MOIRÉ

Amplitude modulated (AM) screening in which images have dots of varying sizes with equal spaces between centers imposes stringent limitations on image orientation. Angles of superimposing images, as in process color printing, are very critical. Angles of 30° between superimposed color separations produce minimum interference patterns called *moiré* (pronounced mwa–ray). Minimum moiré patterns are the rosettes seen in the *magnified dot pattern* illustration on page 90. Since contact screens consist of dot patterns at 90° to each other only three images with 30° angles between them can be used before the angles repeat. In four-color printing, two of the colors must be at the same angle or separated by other than 30°. Usually, as yellow is a light color, it is printed at an angle of 15° from two other colors, generally cyan and magenta. Usual screen angles are: black, 45°; magenta, 75°; yellow, 90°; and cyan, 105° (see illustration, page 90). An error as small as 0.1° between screen angles or a slight misregister between colors can cause serious moiré in areas where three and four colors overprint.

MOIRÉ PATTERNS

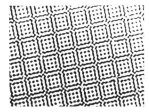
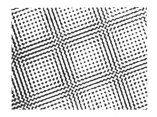

In electronic dot generation, it is difficult to obtain exact 15° and 75° screen angles. Two approaches to this problem have been used. One uses screening algorithms that produce screen angles of 18.4° and 71.6° (called rational tangent angles) for the cyan and magenta and different screen rulings for the yellow and black. The other approach uses screening algorithms that produce angles with errors less than 0.05° for the 15° and 75° angles (called irrational tangent angles) and the same screen rulings for all the other colors.

Frequency modulated (FM also called *stochastic*) screened images, which have very small dots of the same size or varying sizes (second order FM) and varying distances between dots (see page 67), corresponding to the tone values are independent of screen angles and screen rulings. This is also true of continuous-tone printing processes like collotype and screenless lithography. Therefore they are free of moiré from these sources. Also, very fine screens from 300 to 600 lpi, as used for waterless and Hi-Fi printing, produce such small rosettes that moiré is practically invisible.

Moiré patterns can also be caused by improper transfer of ink, a condition called *poor ink trapping*. Proper ink trapping occurs when the same amount of ink transfers to both the previously printed and unprinted areas of the paper. If inks are not formulated to trap properly, the print will have weak overprint colors (red, green, and blue) and moiré.

COLOR PROOFING

The main purposes of proofing are to see (1) if all the image elements are present, if they fit and are in the right color, and (2) how the job will look when it is printed. (1) is done on computer monitors and are called *soft proofs*; (2) is done on proofing systems and are referred to as *hard proofs*. Visual analysis indicates if and where corrections are necessary. After correction, most images are reproofed and shown to the customer for approval. A third proof is usually made when the job is finally assembled and ready for platemaking. With the exception of some magazine advertising which still uses press proofs, most proofing is done off-press, and there are two types: *analog proofs,* which are made using films and analog proofing systems, and *digital proofs,* which are made from digital files and digital proofing systems.

Analog proofs are of five types: *overlay* for internal quality control; *laminate, peel-apart,* and *ink-on-paper* for contract proofing, and *photographic* for checking position, color distribution, register, trapping and other quality factors. Many analog proofing systems have difficulty recording the very small (as low as 14 micron) dots used in FM halftones. Also analog proofs defeat the purpose of computer-to imaging systems as they are made from films and computer-to systems do not use films.

Direct digital colorproofs are of two types: *halftone* and *continuous tone* (contone). Digital halftone proofers are expensive but they have good matches to printed colors. Three technologies are in use: *dye diffusion thermal transfer, laser ablation transfer* and *photographic.* Digital Contone proofers are reasonable in cost and systems with good calibration have color matches close to those made by halftone proofers. Contone proofs have become more acceptable as contract proofs with the increasing use of FM screening. Three main technologies for contone proofing are *ink-jet, thermal dye sublimation* and *thermal transfer.*

Interactive remote colorproofing Contone digital proofers make possible *interactive remote color proofing* in which printers and customers use identical calibrated color proofers viewing the same digital files at different locations. A new soft color proofing system has been developed using flat panel displays with cyan, magenta and yellow (CMY) pigments in place of the red, green and blue (RGB) displays of computer monitors. A major problem with computer soft proofs is that the RGB proofs do not match the colors of ink on paper. The flat panel displays with CMY *liquid crystal* pixels produce images with very close spectral matches to ink on paper. These displays are being developed for use as *interactive remote color proofing.*

IMAGE ASSEMBLY AND IMPOSITION

The terms *image assembly* and *imposition* are similar. Image assembly is called *stripping* in lithography. Imposition originated in letterpress. All three terms refer to the assembly of image elements to create pages and signature forms for printing.

The important fact to be stressed is that both imposition and stripping depend on a plan for the assembly of the images for printing. In the photomechanical processes, the plan is often referred to as a press layout (not to be confused with an art layout). The term imposition is still used in letterpress.

This layout, or imposition, is needed at the planning stage when a number of pages are to be printed in the same form. This is to ensure that when the sheet is printed, folded and trimmed, the pages will appear in the proper sequence. Some presses print layouts of single pages, others print layouts of pages for magazines or catalogs; and still others print layouts of multiples of the same subjects such as labels and packaging. In any case, all material for printing must be planned to produce products in the proper position for finishing or binding.

RETOUCHING, OPAQUING AND REGISTERING

After negatives or positives have been made, the individual films must be properly prepared before they can be assembled to make plates. For conventional gravure, the continuous-tone negatives and positives are retouched so they have the proper tone values for printing. For halftone gravure, letterpress and lithography, and for line drawings and text, the films are trimmed; marks for register, margins, bleeds, etc. are inserted; and films are opaqued to remove pinholes and other flaws.

Pin Register Systems

In addition to register marks, films are most commonly registered by the use of pin register systems. These consist of devices for punching holes and slots in films and copy, and pins which fit in the holes or slots so the copy or several pieces of film can be held to ensure exposure or placement in the correct position. Pin bars, in which pins are spaced at the same distance as the holes in the film, are used for making multiple exposures on films or plates. Pin register devices are important for color reproduction and are used throughout the prepress process from the original copy to image and platesetters, and mounting the plates on the press.

FILM ASSEMBLY OR STRIPPING

Layouts or impositions will vary among printers because of variations in the type and size of printing and binding equipment. For special jobs, the layout should be prepared by, or checked with, the bindery to make sure it can be processed in the folding or other finishing equipment. Certain basic rules are followed. After page size has been determined, ⅛″ to ¼″ is added on the top, side and bottom for trim. If the page numbers or folios are not already on the film, they are inserted at this point. On larger signatures, the center or gutter margin is varied according to the position of the page in the signature and the bulk of the paper. This is called *shingling*.

Layouts for Printing

Sheetfed There are three basic types of arrangement of pages for sheetfed printing: (1) *sheetwise,* (2) *work-and-turn* and (3) *work-and-tumble*. In sheetwise layouts, different pages are printed on each side of the sheet. It is used when the number of printed pages is large enough to fill both sides of the sheet as in signatures for books, magazines and catalogs.

In both *work-and-turn* and *work-and-tumble* layouts, the front and back of the sheet are printed on the same form, and there are two finished units to the sheet. In work-and-turn, after the first side is printed, the sheet is turned over from left to right for printing of the second side. The same gripper edge is used for printing both sides; the side guide changes. In work-and-tumble, after the first side is printed, the sheet is turned over from gripper edge to back for the printing of the second side; the side guide remains the same. After printing, the sheet is cut in half for folding. Changing the gripper edge and/or side guide can cause problems in register unless the paper is accurately squared and trimmed before printing.

The layouts shown on the following pages illustrate the most common impositions for 4, 6, 8 and 16-page forms printed on sheet-fed presses. 'X' indicates the gripper edge.

Web-fed layouts for web-fed presses pose fewer problems. The paper on web presses is printed as a continuous web, fed into the folder, and processed while still under film control by the press. Imposition planning for web registration is greatly simplified. The basic dimensions of the printed sheet are fixed by the web width and by the cylinder circumference, or *repeat length*; and registration of the web is handled automatically during printing.

CONVENTIONAL PREPRESS WORKFLOWS

Conventional prepress workflow starts with a layout and preparation of copy to make a pasteup of the copy for photography that produces the films for retouching and color correction to be assembled in the page layout. On large jobs copy is often sent in piecemeal. It must be keyed and either processed and stored, or filed until all the copy to complete a form is received. Much time is wasted looking for misplaced, misfiled, or lost copy. Line films are made of the text and black and white (B/W) line illustrations. Halftone films are made of the B/W illustrations on cameras. Color illustrations can be transparencies, prints or artwork. These are color separated on electronic scanners, output to films and processed. With the exception of electronic scanning, most of the steps in the conventional prepress workflow are manual operations that are very skill-, time- and cost-intensive.

After the B/W films are made they are checked for information and size. The color separation films are proofed to check for size, information and color errors. If there are errors in the B/W films they are usually rephotographed. If the errors in the color separations are minor they are corrected by manual or photographic dot etching; if serious they are rescanned. After reproofing, all the films are assembled into the page layout and proofed again for approval by the customer.

Following approval, the pages or other copy are assembled on the layout of the signature for printing which is called the *imposition.* Register, trim and other marks are added to assure register in printing and performance in bindery after printing. This layout which is called a *flat* is usually made with negatives taped on a sheet of colored plastic or paper (goldenrod). Windows are cut in the plastic or goldenrod in the image areas to expose the plate. Some complicated jobs like maps may require three or more separate exposures on the same plate. These are made using pin register systems. If the plates are to be made from positives the films are laid out in position on a large plate-size film. All the information must be on the film as multiple exposures are almost impossible to make with positive films.

DIGITAL PREPRESS WORKFLOWS

Most of the operations performed and described in the *Imaging for Graphic Arts* and *Image Assembly and Imposition* sections are manual, time and skill intensive and expensive operations. As explained on page 60 in the *Copy and Art Preparation* section, almost all aspects of copy preparation and processing are performed by personal computers and software programs. Operations involving films are being eliminated by the use of computer-to systems. In fact most of the prepress operations have been replaced by digital imaging hardware and software. Digital workflows for most of these operations are described in the section on *Digital Prepress (pages 91-114).*

TYPICAL FILM OR PLATE SHEETFED IMPOSITIONS

FOUR PAGE FOLDER
Work-and-Turn

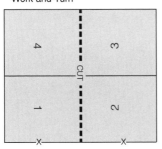

SIX PAGE FOLDER
Work-and-Tumble

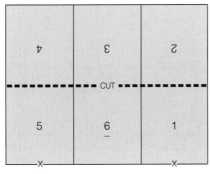

EIGHT PAGE FORM
Work-and-Turn, one parallel, one right angle fold

SIXTEEN PAGE FORM
Work-and-Turn
three right angle folds

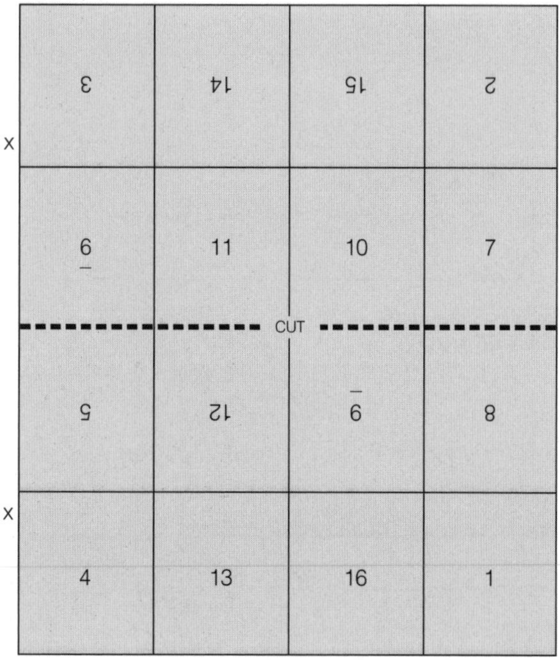

3	14	15	2
6	11	10	7
5	12	9	8
4	13	16	1

X — Press Gripper

The following 12 pages illustrate many photomechanical treatments in both black-and-white and color. Included are various types and screenings of halftones, the principle and demonstration of light and color, four-color process printing, and color process charts. The text is continued on page 91.

GRADATION AND MAGNIFICATION OF TONES

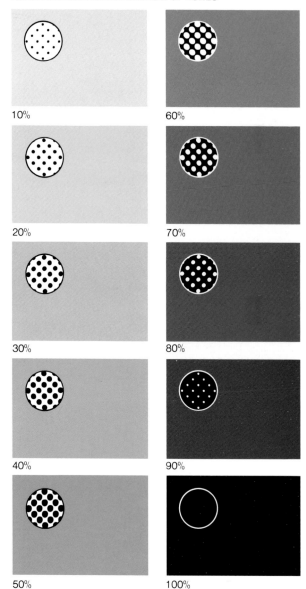

10%

20%

30%

40%

50%

60%

70%

80%

90%

100%

HALFTONE SCREENS

65 LINE SCREEN

100 LINE SCREEN

150 LINE SCREEN

TYPES OF HALFTONES

HIGH CONTRAST HALFTONE

VIGNETTE HALFTONE

OUTLINE HALFTONE

DUOTONES

STANDARD DUOTONE

DUOTONE EFFECT

TINTS AND TYPE

White Type Dropout	White Type Dropout	White Type Dropout
Black Type Overprint	Black Type Overprint	Black Type Overprint

50% COLOR   30% COLOR   10% COLOR

BLACK HALFTONE OVER COLOR

White Type Dropout	White Type Dropout	White Type Dropout
Black Type Overprint	Black Type Overprint	Black Type Overprint

50% COLOR   30% COLOR   10% COLOR

BLACK HALFTONE OVER COLOR

50% BLACK   30% BLACK   10% BLACK

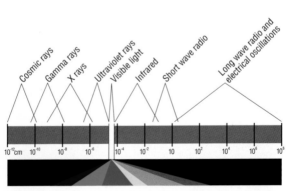

Electromagnetic spectrum showing the narrow band of visible light and color.

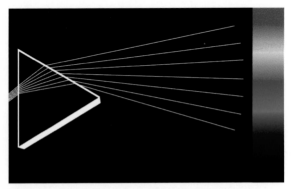

Dispersion of white light into spectral colors by passing through a prism or raindrops (rainbow).

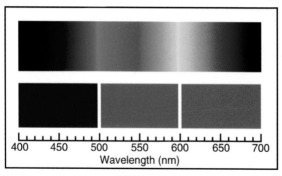

The three-color theory – Additive primaries.

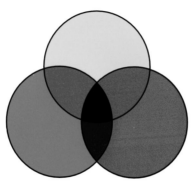

Additive color combinations using light from three projectors covered with broad band filters.

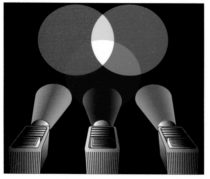

Subtractive color combinations using printing inks.

FOUR-COLOR PROCESS PRINTING

BLUE FILTER / YELLOW PRINTER

GREEN FILTER / MAGENTA PRINTER

RED FILTER / CYAN PRINTER

MODIFIED FILTER / BLACK PRINTER

ROTATION OF COLORS

YELLOW

YELLOW & MAGENTA

YELLOW, MAGENTA & CYAN

YELLOW, MAGENTA, CYAN & BLACK

TWO-COLOR PROCESS CHARTS

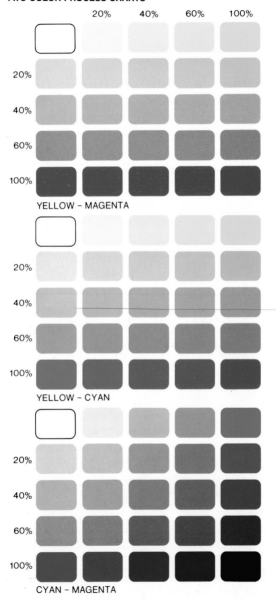

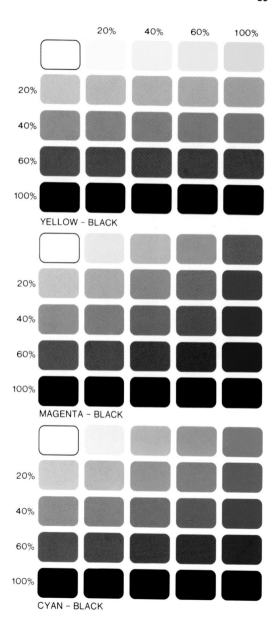

	20%	40%	60%	100%

YELLOW – BLACK

MAGENTA – BLACK

CYAN – BLACK

FOUR-COLOR PROCESS MAGNIFIED DOT PATTERN

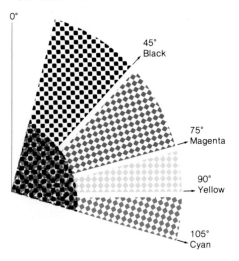

FOUR-COLOR PROCESS HALFTONE SCREEN ANGLES

0°

45°
Black

75°
→ Magenta

90°
→ Yellow

105°
Cyan

DIGITAL PREPRESS

As described in preceding sections, the conventional prepress operations of typographic imaging, copy preparation, photography and film assembly involve intricate manual operations which are very time consuming and cost intensive, and require highly skilled, expensive professionals. These manual operations have been the most serious bottlenecks to production, which is why most new developments in the industry have been targeted at these prepress areas. Although the industry has been faced with these ills for a long time, the means for curing them were not available until the development of color digital image processing and automated workflow systems. Many of these systems have been mentioned in the preceding sections as means of improving productivity, quality and efficiency. This section will discuss the systems in detail and describe the digital workflow that has evolved and is replacing conventional operations.

DIGITAL IMAGE PROCESSING

The first graphic arts processes to be computerized were two of the most important steps in the hierarchy of prepress operations. They were typesetting and color separation by electronic scanning.

Typesetting

Digital typesetting started as *photomechanical* typesetting in 1949 on a converted linecasting machine. It was followed by *electromechanical* typesetting in 1954 using optical projection of type characters on rotating font discs, then *electronic typesetting* in the 1970s with the introduction of the *video display terminal* (VDT) and *optical character recognition* (OCR). It became the *plain paper digital typesetter* in 1985 that developed into the *imagesetter* in 1988. This transformation of typesetting is described in the section on *Type and Typographical Imaging (page 46).*

Scanners

The scanner is one of the most important elements in digital and color electronic publishing technology. Without it, there has been no way to capture and input analog pictures and art. This is becoming academic, as photography is being computerized with the introduction of digital cameras and art is being created electronically. To reproduce a slide transparency, photoprint, painting or other piece of artwork, an electronic scanner has been needed to digitize the image and to convert it to spots that form halftone dots. Scanners are described on page 70.

The first commercial electronic scanner was introduced in 1950. It was a drum-type analog scanner that recorded images as light intensities.

Analog scanners color separate original color transparencies or prints by scanning them with a light beam split into three beams after passing through or being reflected from the original. Each of the light beams went to a *photo multiplier tube* (PMT) photocell covered with a filter corresponding to one of the three additive primaries (RGB). The PMTs converted the varying light intensities to electric currents that were fed to four *analog computers,* one for each color and a fourth for the *black* which was computed from the other three signals *(see illustration).* Note that the output is in a CMYK color space.

PRINCIPLE OF AN ELECTRONIC SCANNER

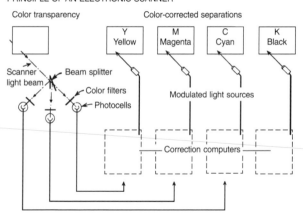

Image manipulations like color corrections, masking, tonal range, UCR, or adjusting for ink, paper and press conditions were performed by the computers, after which the modulated currents were fed to exposing lights to produce the color separation films. Halftones were made by using contact screens in contact with films at output. *Electronic dot generation* was introduced in 1970, and *analog to digital converters* were used for enlargements and reductions.

Digital scanners were introduced in 1982. They are of two types: *drum* and *flatbed.*

Drum scanners are similar to the scanning half of analog scanners. They use similar beam splitters, PMT sensors and RGB filters but process the scans on digital computers. They have high resolutions of 8 to 12 bits per pixel and dynamic ranges of 3.5 to 4.0 and while they are expensive, they have been preferred for high resolution imaging.

Flatbed scanners use RGB filters with *charge coupled devices (CCD)* as sensors. Early systems had low resolution and dynamic ranges of about 2.0. Recent developments using very high resolution (over 6,000 dpi) CCDs have increased the resolution of flatbed scanners to 8 bits per pixel and dynamic ranges to over 3.0, so flatbed scanners can compete favorably with drum scanners. Both flatbed and drum scanners have the ability to perform color adjustments in the original scanning.

Digital scanners record images in an RGB color space and use software programs to convert it to a CMYK color space to produce the color separations suitable for color proofing and printing. *Color calibration* and *color management* systems are used to assure consistent conversion from RGB to CMYK color spaces. The use of digital scanners in production is described in the section on *Imaging for Graphic Arts (page 70)*.

Scanner resolution is an important property of scanners. It is the number of pixels the scanner can resolve and is inversely proportional to the speed of scanning. For *desktop color* lower resolution with higher gray scale is acceptable, but for single color type and line art reproduction at least 600 dots per inch (dpi) but preferably higher are needed to avoid stair stepping or jagged edges.

Optical resolution is the actual fixed resolution of the scanner. It is measured in horizontal and vertical directions, such as 1,200 x 2,400 dpi. In the horizontal direction, resolution is determined by the CCD array. In the vertical direction, resolution is controlled by the sampling rate of the scanning head as it is advanced by a stepper motor that moves in small increments.

Color Management

Color management is a means of ensuring color consistency throughout a printing system from the scanner or digital camera, to computer monitor, to color proofer, to final reproduction on press or printer. Since each device has a different color space and color gamut the color must be managed as it progresses from device to device.

A *color space* defines color. Color monitors which produce colors by exciting red, green and blue phosphors, use an RGB color space. Imagesetters and printing presses use CMYK color spaces. *Color gamut* is the range of color a device can produce. A 24-bit color is capable of 16 million colors but an offset printing press may only be capable of about 5,000 colors.

Each component in a color system works within its own color space and gamut, producing color that is device specific. The scanner represents color with a certain set of data. When that color is displayed on a screen, it is represented by another set of data and appears different. When you print the color on a printer, other hues appear.

Color management translates these device-specific colors into a common set of colors to assure predictable color reproduction using: (1) a standard color-definition mode, such as the CIE color space; (2) profiles that represent the color characteristics of different devices; and (3) transformation algorithms that translate colors from device to device.

Color Electronic Prepress Systems (CEPS)

Early developments in digital imaging were *color electronic prepress systems* (CEPS). These were high-end systems introduced in 1979 and used in the decade of the 1980s that attempted to perform all the steps from the original copy to the press plate in one system of integrated units. They consisted of a high-end drum scanner, or interface with one; a digitizing tablet with cursor or tracer pen; computer storage, software, a high-resolution video monitor for local and area correction and page makeup manipulation; and an output device to record the images of the completed pages or complete plate imposition on film or a printing plate *(see illustration)*.

These were device-dependent systems which used proprietary hardware and software; were very expensive to buy,

ELECTRONIC PREPRESS SYSTEM

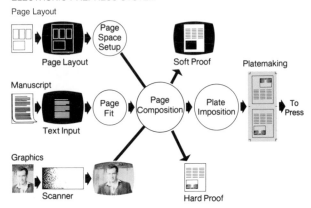

install and operate; and were not interfaceable or interactive with other CEPS systems. In addition, text was not editable and was handled as image elements.

DESKTOP PUBLISHING

Device-independent digital imaging systems began to enter the graphic arts in the 1980s. They use personal computers and off-the-shelf hardware and software that can perform many of the prepress functions of color reproduction. These systems were popularly called *desktop publishing* or *desktop prepress*. The desktop computer was adapted to the printing industry as early as 1982 when the IBM personal computer was programmed to run crude typesetting and pagination programs. Because early PCs were text-oriented, they were integrated into multi-terminal systems for editorial, typographic and pagination functions.

The Apple Macintosh, introduced in 1984, was the first desktop computer to make typography an integral part of its operating system and to offer increased levels of graphics handling. In 1985, Apple combined the Macintosh desktop computer, a laser printer, a page layout program and a typographic page description language into a system that essentially covered the creation and production of typographic pages with desktop devices. For lack of a better term, desktop publishing has come to describe the print production process.

The desktop process is used to (1) output pages from laser printers or high quality imagesetters; (2) produce completely imposed printing plates; and (3) image directly on press. This is possible because of software and page description languages that standardize communication between the computer, its application software and the output device.

Software ties it all together . . .

The desktop publishing revolution has been the result of readily available software that performs the many tasks, usually in an automated fashion, that take documents from creative professionals and pass them on to prepress and printing professionals.

Creative Editorial people use personal computers with word processing software to compose and edit text. They sometimes format the text with interactive or tagged styles using the mouse or keyboard. Illustrators use drawing and painting programs to create line art and business graphics. Photographic professionals use image manipulation programs to scan photographs and modify them electronically. Art and graphic design professionals use desktop computers with page layout software to assemble pages and documents by importing text, line art and photos. In many cases these functions overlap, and it is not uncommon to see a single creative person performing most of these functions. Network software allows files to be transferred from computer to computer either in one office or over telephone lines to multiple offices.

Each program runs in a computer to allow users to perform required tasks. The programs are more or less transparent to the user with *icons* or on-screen pictures that denote certain tasks. Users click on or drag over these icons to activate the function.

Creative workflow (1) author text; (2) edit text; (3) create art and graphics; (4) scan and manipulate images; (5) design and lay out pages; (6) assemble page elements; (7) pass files over networks; (8) proof pages; (9) record and transmit files to prepress.

Prepress The digital document arrives at prepress and is input to their computers. The files are checked to make certain that fonts, images and settings are correct. Preflighting software aids in this task by analyzing files and alerting operators. Specialized software for trapping, color management and imposition may be applied as necessary. The document is then proofed on digital proofers and sent to the client for approval. After any changes are made to the computer file, the document is sent over the network to a RIP (raster image processor), described on page 107 which is connected to an output device. That output device might output film, plate or paper. Other than the marking engines that print, proof and expose film or plates, almost all devices in use today are actually computers running one or a number of specialized programs to link, process and output information in some format. The user interfaces with the software through the graphical user interface, so that functions may be performed via dialog box, menu or key command selections.

PART OF A GRAPHICAL USER INTERFACE

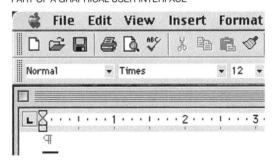

The *graphical user interface* (GUI) is the way information is presented on the monitor. It consists of small pictures (icons), menus and functions that make the system easy to use. The user clicks on icons to select files or programs, or points and clicks at menu items or fill in dialog boxes. Illustrated is part of a typical graphical user interface.

Prepress workflows (1) scan high resolution images; (2) preflight files for fonts, images and settings; (3) trap colors; (4) apply color management; (5) proof pages for color, imposition; (6) impose pages for signatures; (7) direct files to servers and RIPs; (8) output film, plate, paper.

Workstations

There are many brands of desktop computers, from Intel-based versions using Microsoft Windows to the Apple Macintosh, to high-level systems using the UNIX operating system, to proprietary systems. The computer hardware and operating system on which the software runs is called the *platform*, and the combination of the hardware/operating system, monitor, input device and peripherals is called the *workstation*.

The operating system interconnects all parts of the workstation and provides an infrastructure for the application software. The GUI determines how you see and operate from the screen. An application is any program that performs a function such as word processing, database, spreadsheet, page layout, color retouching and many others.

Prepress Application Software

Application software can be categorized as follows:

- **Word processing** programs allow the input, editing, formatting and printout of pages of text.

- **Spreadsheet** programs allow columns and rows of numbers and other data to be organized, formulated and mathematically constructed.

- **Database** programs allow the storage of records (individual items which can be text or images) with fields of data to permit sorting and reorganization for list and file management.

- **Type manipulation** programs distort and manipulate type to create special effects, graphic elements and logos.

- **Image manipulation** programs allow creative modifications of color and images to produce electronic art.

- **Page layout** programs assemble type, graphics and pictures into page form.

- **Synthetic art** programs allow the creation of art and illustration with sophisticated drawing tools.

- **Imposition** programs take page files and position them on film or plates exposed by large-sheet imagesetters to produce imposed plate-ready film flats or plates.

- **Color retouching** software programs can make changes in color images scanned into the system, integrating other images and modifying color.

- **Pagination** is the process of assembling pages of text, graphics and photos into final form. It is the essential step in electronic prepress and for evolving technologies such as on-demand printing and publishing.

- **Batch** pagination requires coded instructions interspersed with the text and other elements. This set of data is then processed through a computer program that responds to the instructions and formats the information into the final page automatically.

- **Interactive** pagination is essentially done by the operators. They create page grids and flow text and other elements onto the page, positioning them as required. This process takes place on-screen where full representation of the type and images is essential for proper positioning. This is the most common approach.

There are application programs for many other areas as well.

Open Prepress Interface

OPI (Open Prepress Interface) is a method for replacing high-resolution images with a placeholder image, called a viewfile, proxy image or FPO image, to minimize the handling of large data files. OPI-compliant publishing programs let you lay out pages as usual, using text, line art, charts, graphics and other images, but substitute a small viewfile with OPI comments for high-resolution images to make your working files smaller and more manageable. This maximizes the productivity of workstations and minimizes time as files are transferred on the network.

As the page is being assembled and proofed, the high-resolution images can be edited as necessary and stored on a server. When both the layout and the images are ready, the high-resolution data can be merged back into the page, in accordance with the positioning instructions contained in the OPI comments. OPI software generates the viewfiles and automates the process of merging the image data back into the page.

OPI is one of the most common forms of image replacement. Other methods are DCS (Desktop Color System) and APR (Automatic Picture Replacement).

Electronic Editing

The video screen or monitor has become the pervasive tool of modern text and graphics technology. It is the most visible part of word processing, personal computers, all levels of prepress and electronic publishing and data processing.

All activity on the screen takes place at the position indicated by the *cursor,* a blip of light that can be moved. Cursor control keys usually have arrows for up, down, left and right movement. The mouse or track pad are the primary controllers of the cursor and on-screen activity.

The data displayed on the screen represents a part of the *file* that is stored in memory. To see a section of the memory data on the screen, a technique called *scrolling* is used to move the memory data up and down to bring the required section onto the screen. In effect, scrolling is like winding and unwinding a scroll.

An alternate approach to scrolling is the *next screen* or *next page* technique which allows the operator to bring the next *screenful* or page of data to the screen in one block.

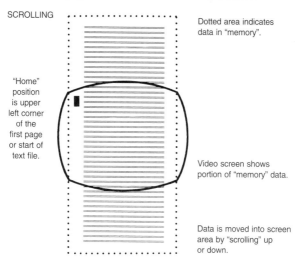

SCROLLING

Dotted area indicates data in "memory".

"Home" position is upper left corner of the first page or start of text file.

Video screen shows portion of "memory" data.

Data is moved into screen area by "scrolling" up or down.

At the cursor position, the operator may add, delete or change a character. Deleted characters appear to vanish as all characters on the right move one character position to the left to close up the hole from the deleted character. In addition, the reverse occurs as each character moves to the right to make room for the new character. A change is usually a substitution of one character for another.

Words that do not fit on a line are automatically moved to the next line. This is called *full word wrap-around* or *word wrap*. Word spaces may be blank or a symbol may be used to indicate their presence.

Functions that are performed on the screen within moments of the operator's request are called *foreground* functions. Functions that must take place off-screen are called *background* functions. Electronic editing is always a foreground approach.

To see the files that are stored on the system's magnetic memory the operator requests a *file directory.* This shows the names of the stored files with additional information, such as the number of characters that are stored. The operator can request a file by calling for it by name and it then appears on the screen with the beginning of the file at the cursor's home position. The operator can also file the material back to memory which records it on the rigid disk or other medium.

File management is the storage and retrieval of information.

WYSIWYG

This acronym for *What You See Is What You Get* refers to the images seen on the screen of a monitor. Traditional PCs display typewriter-like characters, but newer systems display type and graphics at higher quality levels. The screen image is generated from a graphics card that controls the placement and illumination of screen pixels.

High-resolution monitors are over 72 dpi. Measured diagonally, a 19″ monitor has an image of about 11″x14″.

The goal has been to make the screen image look as close to the final printed image as possible; hence, what you see on the screen is what you will get at printout. However, screen resolution is very coarse (72 dpi) compared to printer resolution. This is especially important in dealing with color. Scanners and monitors work with RGB (transmission color), where most printers and all printing presses work with CMYK (reflection color). This requires color management and calibration so that screen images are representative of the final printed product.

TYPES OF DIGITAL GRAPHICS

Digital types of copy and art and their preparation for reproduction are described in this section.

Computer Art

The artist and the printer have long been partners. The result of the art and design process is usually a mechanical with type and black-and-white elements pasted into position. The artist indicated photo position and cropping, dropouts, overprints, tints and other requirements with overlays and marginal notes — called a *keyline* or *mechanical.* The printer would shoot film for all elements and assemble them by stripping. The resultant film would be proofed for approval by the artist and/or client.

Desktop computing and later desktop publishing brought with them new tools for the artist or designer. The most important was the drawing program, which is also called *synthetic art.* This applications program runs in the workstation and enables the artist to use electronic methods for creating shapes, outlines, borders, fills and many other functions. It also allows the artist to trace line art from scanners into Bezier curves, integrate type, pictures and color, and assemble elements into final form.

This electronic art mechanical can be modified in size and color or in any other way. Thus, fine tuning, alternative versions or major changes can be made easily. Special effects, such as graduated tints, 3-dimensional drawing, shadows, illusions of depth and other effects can be created.

The final version can be output to imagesetters, platesetters or digital printers as color-separated film or integrated into pages for later output.

The concept of camera-ready art is evolving into computer-ready art. The artist and designer are called *computer artists.*

Bitmapped Images

Bitmaps are the real-world analog to a mosaic made from tiny colored tiles. In this example, each tile is a *pixel*, or picture element. Lines are depicted as rows of adjoining pixels, and all shapes are outlined and filled with pixels. Changes to a bitmapped "mosaic" image replace tiles of one color with tiles of another color. The illusion of nonexistent colors or grays is achieved by *dithering*, or mixing tiles of the available color or shades of gray. Where there are two rows of tiles, one at 100 percent black and the other at 50 percent black, and there is a need to insert another row of tiles between them, dithering averages or *samples* the adjoining pixels to create the new row at 75 percent black.

Black-and-white bitmaps need one data bit for each pixel. But a single data bit per pixel does not provide enough information to specify a particular color or shade of gray. Images containing 256 grays or colors require eight bits per pixel, and photographic-quality, full-color images require 24 bits per pixel or more to specify one of 16.8 million colors.

NUMBER OF BITS	COLORS
6	64
8	256
10	1,024
12	4,096
24	16.8 million
30	1 billion
36	68 billion

The fixed mosaic nature of bitmaps can produce undesirable results when a part of the image is enlarged or rotated. To move part of a bitmap, tiles must essentially be torn from the surface and moved, leaving a hole (no tiles). Wherever the shifted image lands, it replaces the tiles that were there, erasing the overlaid portion. When a bitmap is printed, the image reproduces exactly as it is stored, in its mosaic tiled form. Its resolution is directly dependent on the bitmap. An image saved at 72 spots (pixels or spots) per inch (spi) will still be 72 spi even if output on a 300 spi printer or a 1200 spi imagesetter. Pixels let you achieve artistic effects that resemble traditional painting with electronic editability. Photo retouching programs work with bitmaps. All pictures are bitmaps.

Object-oriented Art

Object-oriented (also called vector-oriented) graphics are produced by drawing programs and overcome the limitations of bitmaps. Images are composed of mathematically described objects and paths, called *vectors*. Object-oriented applications don't store line strokes as a collection of tiles but rather as a list of drawing instructions.

Objects can be enlarged, reduced, rotated, re-shaped and refilled and the drawing program will redraw them with no loss of quality. Objects can be managed as though each was drawn on a separate transparent sheet and can be stacked and partially hidden by other objects without being permanently erased.

The advantages of object-oriented art extend to the printing phase. Instead of dictating to the printer where each image spot or tile should go, the program mathematically describes the object and lets the printer render the image at the highest resolution possible. Object-oriented graphics are resolution-independent and are excellent choices for illustrations, especially items subject to changes in color, shape, position or orientation.

PAGE DESCRIPTION LANGUAGES (PDL)

Each printout device has its own programming *language*. It is essentially a set of codes that allow it to perform its functions, such as centering copy, changing type size, advancing paper, etc. After pages are assembled on front end systems the screen image is translated into the printer's language via a *printing driver*. If you change the printer, you must change the driver. *PostScript®* is one such standard and is available on imagesetters, platesetters and other printers from low to high resolution.

A digital output system consists of three parts: (1) the *interpreter,* which converts the driver data from the computer into the coding of the printout device; (2) the *raster image processor* (RIP), which organizes font data and creates the page bitmap; and (3) the *marking engine* which actually produces an output image using dots or spots.

Virtually all printout devices are now digital in nature, producing their images with dots. Thus, all output devices now consist of two functions: the RIP for rasterizing input into page bitmaps, and the engine for making marks on paper, film or plates.

PostScript®

In 1985, Adobe Systems introduced PostScript, a page description language for producing typographic pages. This has become the heart of desktop publishing and electronic prepress. It standardized the language that each application program outputs by developing RIPs for many different printers and output devices. PostScript RIPs can be installed in printers with different resolutions. A page file can be sent to a low resolution printer for proofing purposes and to a high resolution printer for final output on photographic paper or film. PostScript, therefore, is device independent.

Like any spoken language, PostScript has a vocabulary. There are commands that control the size of type, the tint of colors or the position of pictures. Virtually every application program running in every desktop computer outputs PostScript and virtually every printer of every type, including some sophisticated prepress and even press systems, accept PostScript-coded files.

To produce the bitmap for a page, PostScript uses a grid coordinate system. *Paths* are established to identify type, graphics or pictures. *Operators* are commands, and they are preceded by *operands*, the variable data needed to perform the command.

Here is an example of some PostScript language code for a black box.

POSTSCRIPT CODE	MEANING
72 612 move to	Go to coordinates for start point
288 0 rlineto	Side 2 of line
-288 0 rlineto	Side 3 of line
closepath	Close Side 4 of line
fill	Fill with black to make solid area
showpage	Output and reset for next page

GRAPHIC FORMATS

Graphic data is stored in a variety of ways called formats, for integration in other application programs. Three of the most popular formats are:

Tagged Image File Format (TIFF)

Tagged Image File Format is the method for storing bitmapped images in various resolutions, gray levels and colors. TIFF was created specifically for storing gray-scale data, and it is a standard format for scanned photographs. TIFF is now considered a standard graphics format, and is called "TIFF/IT".

Encapsulated PostScript (EPS)

Encapsulated PostScript is a popular format for storing vector- or object-oriented artwork. It can also store bitmaps. An EPS file usually contains two versions of the graphic. The main image is a resolution-independent PostScript description for printing on a PostScript device. The second, optional image is a low-resolution, bitmapped preview that can be displayed on-screen. This double-image scheme enables page-layout programs to import, crop and scale high-quality EPS graphics while using the screen version for the user.

EPS files can be resized, distorted or cropped and most programs that perform color separations accept and color separate them. They may also be re-edited.

Pure PostScript File

A PostScript file is a purely coded text-based description of an image, without the displayable screen image that EPS offers. In many applications, you can create a PostScript file and then open the file with any word processor and modify typographic and positioning information (if you know PostScript coding). With a pure PostScript file, you do not need the originating program to print the file. The PostScript file can be sent to a PostScript printer with a download utility.

PORTABLE DOCUMENT FORMAT

This is a document converted to a special coded file that can be displayed and/or printed on almost any personal computer (PC) without the original application program.

PostScript refers to both the page description language that describes the format of a printed page and the interpreter that converts the page description into pixels or bits to control a raster-based output device. As a general-purpose programming language, PostScript contains procedures, variables and control constructs that may cause unpredictability and it cannot be viewed.

Adobe Acrobat, which built upon the base of PostScript, allows users to view and manage documents in an application independent manner and on a computer platform in an independent manner. Some saw Acrobat as a substitute for paper rather than a transportable printing format. But users saw more potential in the Portable Document Format (PDF) than just looking at pages on a screen — they saw the high end of the printing world — direct to film, plate, printer, press — as well as viewable documents for disk and World Wide Web distribution.

Each page of the Acrobat PDF document is independent of the others. In PostScript a "page" is the byproduct of some calculation or procedure. The PDF page approach eliminates the variability of PostScript and provides a foundation for effective digital print production workflow. High-end printing and color controls can be integrated with the PDF file.

When you click the Print button on any publishing program, the page on the screen is converted to a Page Description Language—PostScript code—and it is sent over a cable to a PostScript interpreter or RIP that converts the PDL into a display list of page objects and then a bitmap of pixels to drive the imager. When you distill your page or document into a PDF you are essentially doing the interpretation and display list functions. If your document can be distilled to a PDF, the odds are that it will output reliably.

Acrobat incorporates extended graphics state functions so that color separation can occur more effectively. OPI image replacement can now take place and the PDF can be exported as an EPS for insertion in a page makeup program.

In use, the Acrobat PDF takes a file of unwieldy PostScript code and distills it into a database of objects on a page. It is highly compact with all embedded fonts, pictures and compressed vector objects. It is device and media independent with CMYK and named (spot) color support and color space definitions incorporated. It can be trapped, imposed, converted and is viewable by Macintosh, DOS, Windows, Unix and Web browsers using the free Acrobat Reader.

PDF File Components

- A view file that displays the page as you created it.
- Embedded type: Adobe Type 1 and TrueType.
- Graphic objects: bitmaps and compressed vector images.
- Links for variable forms data.
- Sound, QuickTime, hypertext-like linking.

The PDF embeds all the elements.

Device Independence

The need for printing drivers or printer description files has made printing more device dependent than users would have liked. The PDF provides the best method yet for storing and defining a document for print. The same file can be used for digital proofing, large format (imposed) film imagesetting, computer-to-plate, black-and-white document printer, digital color printer/color press, internet viewable and printable documents, and archivable formats that can be re-constituted into the original objects so pictures can be re-converted for other applications.

PDF Publishing

PDFs can be used to publish in print or non-print form. For instance, a year's publications can be converted to PDFs and recorded on a CD-ROM. Searches can be made through all issues to find keywords and information. PDFs can also be put on the World Wide Web with a password to create an archive of issues from past years. PDFs retain the look and feel of the publication on-screen or printed out on almost any PostScript printer.

Digital Ads

PDF may be an answer for the digital distribution of advertising, one of the last enablers that will allow *computer-to* technologies to be used by magazine, journal and newspaper publishers. Document creators who distill their pages into PDF and graphic services will not have to worry about what is missing — graphics and fonts are embedded. Trapping can take place in the RIP with imposition. The PDF can support high-end color output from digital color proofers, to film imagesetting, to computer-to-plate, to color presses. TIFF/IT is another method for moving documents around as a set of bits, zeroes and ones, but the PDF file is intelligent — it contains the ability to carry color information, support changes and allow other operations, like trapping and imposition.

RIPS / SERVERS / NETWORKS

Once in electronic or digital form, page document and artwork files can be output on a variety of devices at varying quality levels. Keys to all output devices or systems are RIPs, Servers and Networks.

Raster Image Processor (RIP)

At output, the file is converted into PostScript code, which describes the page, and is sent over the cable (network) to the RIP and the output device. A desktop printer can be used for proofing but a prepress service or printer may be needed to output a film or plate. In this case, the file could be stored on a disk and sent by delivery service. Or, it could be sent by telecommunications. In all cases, the PostScript code must arrive at a RIP in order to be processed for output.

The RIP is the control device for raster-based output. Raster means line. For each line (and there could be from 300 to 3,000 lines to an inch), the laser is either on or off for each possible position. Since almost all output today is based on dots, a RIP is required. The RIP is connected to a marking engine which uses lasers to put dots that create images on a substrate, such as film (imagesetter or imposetter), or plate (CTP) or directly on paper using toner, ink-jet or dye sublimation methods. The end product of the RIP is a bitmap of zeroes and ones that describes where every dot is located. This data is used by the marking engine to turn the laser on or off to place the dots in position. The laser scans across the substrate line by line, turning on or off to make marks (images) on film, plate, paper or any other substrate.

Digital imaging makes direct-to, or computer-to outputs possible.

Servers

A Server is a computer with a large amount of data storage that has shared software and information. A RIP usually has a large disk to store software and pre- and post-ripped files. It is called a *RIP Server.* The RIP should be equipped with OPI functionality. The disk and system associated with the RIP is then also called an *OPI Server.* A Server could fully support Adobe Acrobat with its collection of high-end printing functions. A PDF Server can trap, impose and color manage PDF files.

A network could have all of the above functionality, or some combination; or like most, it could integrate the Server (for the network, image storage, OPI and ripping) and the RIP into one Server, or the Server into one RIP.

Server Integration

A prepress network usually consists of multiple workstations, a file server and a printer or printers connected by network cables.

Sometimes a print spooler is added so that files destined for the printer are accepted by the spooler to release the workstation's application. The spooler then queues the job or jobs based on prioritization and sends them to the printer. The difference between a print spooler and a print server is based on the amount of time files remain on the disk. A spooler may delete files sooner than a server.

The printer on a prepress network is usually a PostScript-based printer and therefore has a RIP. Most desktop printers have the RIP inside the printer . . .

. . . while imagesetters and platesetters and proofers have the RIP as a separate unit, connected by a cable. Today, every raster-based output device must have a Raster Image Processor.

A server may be added to the output function to store ripped files for later printout of the entire file of one or more of the color separations. This may be called a print server or a print spooler. We are going to call it a RIP server because it serves the RIP. At one time there was a RIP for each and every output device. Today, the trend is to try to RIP once and output many times. So the RIP server functions as a print spooler/server, holding files for later printout or archiving.

The RIP server and the RIP may be integrated into one unit.

An OPI server could be added to the system for automatic picture replacement.

Or the OPI server and the RIP server could be integrated while the RIP is a separate unit . . .

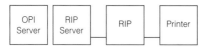

. . . or the OPI server, RIP server and RIP could be integrated into one unit with one or more printers.

When the functions of OPI, RIP and print server are integrated into one Server, we are going to call it a Super Server. It then is networked with the file server and the printer or printers.

To the Super Server we add the final ingredient of PDF handling, trapping and imposition. The final system now takes form as a *PDF Server.*

Networks

A Local Area Network (LAN) is a collection of hardware, software and users brought together so as to allow them to cooperate in a fully integrated environment. A LAN typically covers a limited geographical area, measured in meters rather than in kilometers. LANs can cover the linking of two to several hundred users spanning a single office, to one or more departments spanning several floors of a building or spanning an entire site. LANs usually complement Wide Area Networks (WANs) to extend this environment to interconnect or bridge LANs locally or across great distances to form larger networks.

OUTPUT ALTERNATIVES

Outputs for *Computer-to* Technologies

Following are computer-to technologies in use or development, and systems used for imaging them:

(1) **Computer-to-film** Raster-based imagesetters, both capstan and drum oriented.

(2) **Computer-to-imposed film** Imposetters, 4-up and larger film format.

(3) **Computer-to-polyester plate** Imagesetters or imposetters, outputting film or plate, or dedicated devices that only output polyester plates.

(4) **Computer-to-metal plate** Platesetters using internal or external drum, flatbed or other approaches that expose silver-based, photopolymer, hybrid or thermal plates.

(5) **Computer-to-plate on press** Interface to technology presently integrated on presses that images the plate on the press.

(6) **Computer-to-plate image cylinder** Interface to developmental technology that involves spraying a substance on a press cylinder and imaging it with thermal lasers.

(7) **Computer-to-electronic printer** Interface to all digital printout devices, monochrome or color, using electrophotography, ink-jet or other technologies.

(8) **Computer-to-electronic color press** Interface to special high-end higher-speed color printing systems.

(1) (2) (3) are discussed in this section. (4) (5) (6) are discussed in the section on *Platemaking (page 122)*, (7) (8) are discussed in the section on *Digital Presses (page 147)*.

Image-Impo Platesetters

Typesetter-to-imagesetter The photographic typesetter which was born in 1949 could only set type. By 1988, every supplier of phototypesetters had converted to the production of photo imagesetters, which could output graphics and pictures in addition to type. The ability to output all page elements in position reduced traditional camera and stripping operations, paving the way toward computer-to technologies.

Like phototypesetters, imagesetters are usually roll-fed or capstan, which means photo material is stored in a light-tight cassette or box and then pulled across the imaging area into a receiving cassette.

Drum-based imagesetters are preferred for color reproduction because they produce more accuracy and repeatability of images for each of the four films. These devices use rolls or sheets of film capable of 1- to 8-up standard pages. Two sets of four-color separations can be output on one sheet of film, or eight sheets can be output of 8-up pages, two for each CMYK

A DRUM IMAGESETTER

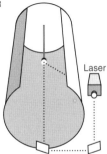

Laser

color. Drum-based imagesetters are either internal drum, where the film is positioned inside the drum and the imaging system moves, or external drum systems, where the film is positioned on the outside of the drum and the drum rotates while the imaging system remains stationary.

Imposetter Imagesetters traditionally used photo material 12″ to 13″ wide (with some 18″ wide) which made it possible to output two standard pages side by side, or an 11″x17″ spread. The 18″ imagesetters made it possible to output four standard pages. With the proper imposition, the film output would have pages in the proper position for platemaking and printing.

Because of the need for 8-up flats some drum-based imagesetters increased their sheet size to 30″x 40″ and even 48″x 60″. Combined with imposition software, they produce film negatives ready for platemaking. Thus, they are called imposetters.

Platesetter The film is replaced by plate material. Although imagesetters can output polyester plates, the need for imposition and color capability delayed computer-to-plate technology. The metal-based platesetter is the engine that makes commercial *computer-to-plate* (CTP) systems possible. Three types of systems are in use: external drum, internal drum and flatbed. (Laser wavelengths vary from 365 nm UV to 1064 nm-IR-YAG.) UV lasers (365 nm) and Violet (405-410 nm) can be used with some high-speed photopolymer plates. Infrared laser types (above 760 nm) are capable of imaging thermal plates. The plate types are described in the section on *Platemaking*.

All other outputs for digital files are described in sections on *Platemaking (page 115)* and *Digital Presses (page 147)*.

DIGITAL WORKFLOWS

Now that the Portable Document Format has become a predictable, platform-independent, page-independent entity which solves many of the limitations of PostScript, we must tackle other print-related workflow issues. Issues such as trapping, imposition, OPI and digital proofing play an important role in the success of PDF as a link to the completely digital workflow for the high-end printing and publishing world.

Each of these functions has evolved from manual techniques to computer operation with manual intervention and now totally automated approaches. For a long time each function often required a different server to perform a function or functions. The trend today is toward more comprehensive and cohesive systems that are transparent to the user.

Preflighting covers all the information and skills needed to quickly, accurately and effectively determine if all job components required for output, and/or proofing, are met before a project enters the production workflow, including:

• Font issues
• Completeness of electronic mechanicals
• Missing page elements
• Linked image files

Trapping is the intentional overlapping of colors to compensate for register errors on the press. It can be accomplished at the application program level, at a separate computer with dedicated trapping software, or at the RIP equipped with trapping functionality.

Automating the analysis process for a PostScript file has two approaches. The first approach parses the PostScript code to identify the logical objects on the page (text strings, tint blocks, contone images, line drawings and other objects) and their relationships to each other. It works with geometric properties, and is called the vector method.

The second approach rasterizes the file and then analyzes the resulting raster image to determine where colors are adjacent — called the raster method. Hybrid systems use a combination of vector and raster technologies.

Imposition Due to the prevalence of large-format imagesetters and platesetters, many users are turning to imposition programs for workflow automation. Some imposition functions:

- Standard and custom imposition layouts for sheet or web printing.
- Form, file and page-level positioning and rotation, with verso/recto page controls.
- Enhanced shingling and bottling controls.
- Customizable page and sheet marks.
- On-screen preview of press sheets with all marks and pages in place and proportion.
- Support for pin-registration systems with full control over form and sheet position.
- Accommodate all binding methods.

OPI workflows can improve prepress system performance by reducing the amount of data that workstations and networks must carry and process. An OPI Server keeps high resolution graphics stored until imagesetter or printer or platemaker output time, and creates a low resolution "view file" for applications to work with.

An OPI Server adds the ability to "OPI-Publish" TIFF images from the Server database. For each high-resolution TIFF image which is "OPI-Published," a view file (a low-resolution of the same image) is made available. When users of OPI-compatible applications need TIFF graphics, they can use these view files instead of the actual high-resolution graphics.

Digital proofing Moving to dotlessness may be the theme in the color proofing area. The contract proof (the verification proof that the client, service bureau and printer agree will be the standard for color and quality) issue is moving toward an uneasy acceptance. Most contract proofs have used film-based technology, but with the increasing move to all-digital workflows, and as computer-to-plate systems avoid film entirely, film-based proofing is being replaced by proofs from digital data.

A contract proof has traditionally shown the exact halftone dot structure so that potential printing problems like moiré can be avoided. Some digital proofing systems reproduce halftone dots, and some do not.

Interactive remote proofing is a major trend in which an ink-jet or dye sublimation color proofer is physically installed in a customer location. PDFs from the customer are sent to the prepress or printing service and processed. The Server system prepares a version of the PDF that calibrates to the eventual reproduction device and the files are returned to the customer for proofing printout. Since the PDF files are compressed, they can be sent via telecommunications lines and discussed interactively.

New Servers and RIPs will automatically direct PDFs to queues based on the customer, reproduction device (litho press, flexo press, digital color press, etc.) and convert or modify files as required for each step in the workflow.

More information on digital proofing and interactive remote proofing is in the section on *Imaging for Graphic Arts (page 73)*.

SUMMARY

In the 1960s our industry was moving out of its mechanical phase where jobs went from machine to machine for various functions. Phototypesetting and electronic scanning had been introduced in the 1950s but they were struggling for acceptance. In the 1970s computers entered the industry to automate certain tasks. The Video Display Terminal (VDT) and Optical Character Recognition (OCR) gave phototypesetting the impetus it needed to succeed; electronic dot generation (EDG) lifted scanning out of the doldrums; and during the decade ink-jet printing and computer-to-plate technologies were born. In the 1980s we entered the digital age, which merged the computer and software modules with output technology based on binary spots. The 1990s accelerated the application of digital technologies that have almost completely converted printing from an art to a science.

The result has been the ability to take a digital document and re-format it so that content formatted as a book might be automatically converted to a CD-ROM or transformed into a web page. The essence of digital workflow today is *re-purposability,* the ability to re-direct and re-format content to print or use other media through conversion software. Printing, which for many years was the main source of graphic communications, is now just one of the many outputs of digital imaging technology that supports all the media.

PLATEMAKING

Platemaking, or, more correctly, making the *image carrier* (as some processes like gravure and screen printing use other imaging surfaces), used to be the culminating step in the prepress process where the results of all prepress operations combine to fit the needs of the different printing processes. Platemaking is now more logically part of the press or printing operation. Digital imaging has opened new vistas for platemaking. Instead of being dependent on the prepress operations of film making and assembly, image carriers can now be made directly from digital data without the need for films and their assembly into plate layouts.

The main beneficiaries of this new technology are lithography, gravure and flexography. Each printing process uses a different kind of image carrier which relates to the characteristics of the image to be produced, the type of ink and press used, and the number and speed of impressions to be printed. This section describes the conventional image carriers for each printing process, and the new digital computer-to systems for the different processes when applicable.

METHODS OF PRODUCING PRINTING IMAGE CARRIERS

Image carriers can be made in a number of ways depending on the printing process, length of run, type of press, etc.

Manual image carriers consist of woodcuts and linoleum blocks for letterpress, copperplate or steel-die engravings for intaglio printing, stencils for screen printing, and images drawn on stone by artists to produce original or limited editions of lithographs.

Mechanical image carriers have been used to make duplicate printing plates for letterpress and pantograph engravings, and engravings made with geometric lathes for intaglio printing.

Electromechanical engraving (EME) machines are used extensively for making gravure cylinders, including computer-to-cylinders.

Photomechanical platemaking has been the most universally-used method of platemaking for lithography, letterpress, flexography and conventional gravure. It uses light-sensitive coatings on which images are produced by image-wise light exposure through photographic films and processed according to the requirements of the printing method used.

Electrophotographic imaging uses xerographic principles, photoconductors and toners to make lithographic plates for printing on duplicating presses and digital printing systems.

Thermal imaging uses lasers and new thermal polymers with threshold binary imaging, thermal ablation transfer and new thermal crosslinking technologies.

Digital plates are produced by computer-to-plate (CTP) technologies using lasers in platesetters driven by digital data from prepress, desktop publishing systems or digital data files.

PHOTOMECHANICS

The light-sensitive coatings used in photomechanical processes change in physical properties after exposure to visible radiation (light). Usually the exposed areas change in solubility in water or other solutions. Means of imaging printing plates other than visible light exposure are UV radiation, lasers, laser diodes and thermal energy (heat).

When photomechanics was first introduced, typical coatings were natural organic substances such as asphalt and shellac, followed by natural organic colloids like albumin, casein and gum arabic sensitized by ammonium or potassium bichromate. These coatings are obsolete because of the loss of sensitivity of bichromated coatings on standing and the rapid decay of the latent image after exposure, as well as its use resulting in dermatitis and heavy metal toxicity. Bichromated coatings have been replaced mainly by diazo compounds and photopolymers.

The latest plate technologies are CTP systems. These are computer-to-plates made directly from digital files and use plates with silver halide or high-speed photopolymer coatings, hybrid mask-coated plates, new thermal plates using laser ablation, laser ablation transfer or thermal crosslinking technologies and a no-process ink-jet imaging system.

Plate Exposure and Processing

Plate exposure and processing are critical operations in platemaking that have caused uncontrolled variations in the process. Developments in both areas have improved control in platemaking. For exposing negatives or positives on conventional plates, high-intensity pulsed xenon and metal halide lamps are used in one of two methods: (1) *vacuum frame* and (2) *step and repeat*. CTP systems are exposed in special platesetters with lasers compatible with the plate coatings.

Vacuum frame Exposures on plates are made in a vacuum frame if all the negatives are stripped on a single flat or if the same flat is to be exposed two or more times on a plate, in which case pin-register devices are used to make sure that the exposures are made in the proper position. Sometimes two or more flats are exposed on the same plate. This is known as *surprinting* or *multiple burns*. This can be done with negatives, but not with positives. When plates are made from positives, all elements must be combined on one flat before the exposure is made.

When exposing in vacuum frames, it is important that full vacuum is used. Otherwise undercutting of the image occurs which causes dot gain due to light spread on plates made from negatives, and causes dot sharpening on plates made from positives. Newer vacuum frames have computerized exposure programs for different uses such as making plates, proofs or contact prints. In addition, some have compartmented blankets for high-speed vacuum draw down to increase efficiency and assure even exposure, especially of FM screened images.

Step-and-repeat When five or more exposures are to be made on a plate, it is usually more economical and accurate to use a step-and-repeat machine which is designed to produce multiple images of negatives or positives on a printing plate. It consists of a bed for mounting the plate which is usually held by vacuum during exposure; a chase for mounting the film for exposure; a means for traversing the chase accurately in two directions; and a high-intensity lamp-like pulsed xenon or metal halide for exposure.

Some step-and-repeat machines have devices for moving the chase in both directions automatically using computer programs. Some are also equipped with film cassettes so that they operate automatically by rejecting one film after all the exposures with that film have been completed, and picking up the next film for continuing the exposures on the plate. Such automated machines are very useful in packaging and label printing

STEP-AND-REPEAT MACHINE

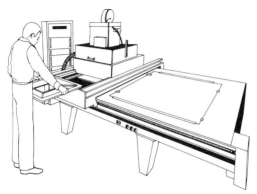

where a number of exposures of different subjects, shapes and sizes are made on a plate. They are also useful in book production where the negatives are programmed in advance for exposure in the proper position on the plate. For label printing, where many different size labels appear on the layout, computer programs are available to produce the optimum layout and schedule for the exposure.

Hybrid step-and-repeat machine Two manufacturers of step-and-repeat machines have added digital imaging heads to one of their step-and-repeat machines so that the systems can make exposures with films and direct digital imaging on the same plate. These are transition systems that allow the combination of digital files with supplied films for publications, packaging and label printing until all film libraries and supplied films are converted to or supplied as digital files. These machines help accelerate the adoption of CTP technologies in these market segments.

Projection systems Another method of step-and-repeat is to use computerized projection imaging systems which are large darkroom cameras with plate-size precision-movable vacuum backs that move in X, Y planes so all exposures are made on the optical axis of the lens. These systems are still in use and supported, but are no longer manufactured.

Platesetters CTP plates are made in platesetters which consist of means for registering plates, a RIP for converting page description files into bitmapped halftone digital files that drive a laser to expose the plate. Three types of platesetters are in use: flatbed, external drum and internal drum. Gas and solid state lasers with wavelengths from 365 nm to 1065 nm are used for imaging the plates. Flatbed platesetters are used for high-speed UV photopolymer plates using 365 nm, 405 nm lasers and polyester-based silver halide plates using 488 nm and 670 nm lasers. Metal base CTP plates use external drum and internal drum platesetters. The plates are stationary in internal drum and rotating on external drum systems during exposure by laser beams.

Choice of platesetter/laser combinations depends on the plate system to be used. High-speed UV photopolymer plates use 375 nm (UV) and 405 violet laser diodes; silver halide, dye-sensitized photopolymer, and hybrid mask-coated plates use visible light lasers in wavelengths 488 nm (blue), 532 nm (green) or 670 nm (red). Thermal plates require infrared lasers in wavelengths 832 nm, 860 nm or 1064 nm. No-process ink-jet plates require special platesetters as they do not use lasers.

Automatic processors Plate processing has been a major source of variation in platemaking. Automatic processors have stabilized platemaking almost as much as they have photography. They process the plate after exposure, gum and dry it. In newspaper printing some processors combine exposure with the processing, gumming and drying, and several include coating as well. Most platesetters have integrated automatic plate processing machines. The latest plate technology, especially for CTP, uses chemical-free and processless plates.

LITHOGRAPHIC PLATES

Lithography is a planographic process using thin metal plates (except for the few litho stones still used by artists) with the image and non-image areas essentially on the same plane. Lithography is based on the principle that oil and water do not mix. A lithographic plate is precoated with a light-sensitive or thermal-sensitive imageable coating and the separation between the image and non-image areas is maintained chemically. The image areas must be oil-base ink receptive and refuse water, and the non-image areas must be water receptive and refuse ink. In reality, ink and water do mix slightly. If they didn't, lithography would not be possible. If they mix too much, there are problems. The wider the difference maintained between the ink receptivity of the image areas and the water receptivity of the non-image areas, the better the plate will be, the easier it will run on the press, and consequently, the better the printing.

Ink receptivity is achieved with inherently oleophilic (oil-loving) resins on the image areas. Water receptivity of the non-image areas is achieved by using metals like aluminum, chromium or stainless steel whose oxides are hydrophilic (water-loving). Water receptivity is maintained in platemaking and storage by treating them with natural and synthetic gums, mainly gum arabic. Almost all lithographic plates use metal bases of grained and anodized aluminum. Most are treated with silicate solutions to protect them from corrosion and help strengthen and retain their water receptivity.

There are two types of lithographic plates: (1) conventional and (2) CTP. Conventional plates have light-sensitive coatings that are exposed to photographic negatives or positives using high-intensity lamps. CTP plates use coatings exposed by digitally-driven lasers.

Conventional Lithographic Plates

Six types are in use: diazo, photopolymer, high-speed UV photopolymer, waterless, bimetal and on-press processed plates. These are imaged by film negatives or positives. Positives produce plates with less dot gain than plates made from negatives and are used extensively for web offset publication printing. Bimetal and some high-speed UV photopolymer plates have better control of dot gain than positive plates.

Diazo plates Diazo coatings are organic compounds that are used to make presensitized plates with a shelf life of about a year, and for wipe-on plates that can be in-plant coated, with a shelf life of one to two weeks. There are negative and positive plates, and several types that can be made from positives or negatives. Diazo presensitized and wipe-on plates are easy to make. Most are made from negatives. Once exposed, they are treated with an emulsion developer which consists of a lacquer and gum-etch in acid solution. As the unexposed diazo is dissolved by the solution, the gum deposits on the non-printing areas ensuring water receptivity and lacquer deposits on the exposed images making them ink receptive. Once developed, the plate is rinsed with water and coated with a protective gum arabic solution. These are known as *additive* plates and can produce runs as long as 150,000. Some diazo plates are pre-lacquered which make them capable of runs up to 250,000. These plates are processed with special developers, originally solvent based, but more recently aqueous based developers have been dominant. These are known as *subtractive* plates. The images on most plates made from positives must be stabilized during processing.

Photopolymer plates Coatings used to make photopolymer plates are organic compounds which are very inert and abrasion resistant, making them capable of longer press runs than diazo coatings (up to 250,000). They are available as negative and positive plates. Photopolymers are different from other sensitizers — they change in molecular size and molecular weight during exposure. This accounts for many of their desirable properties such as long runs, resistance to abrasive wear and increase in wear resistance after baking. Most photopolymer plates can be baked in special ovens after processing. Baked plates have been used to produce runs exceeding one million impressions.

New photopolymer plates have been introduced with very high resolution using new aqueous-processed coating and aluminum graining technology. They are capable of printing 0.5% to 99.5% dots on 150 lpi screened halftones. They have been used for printing 400 lpi screen images and tested on images with screen rulings up to 800 lpi. There are also new photopolymer plates capable of printing over 2,000,000 impressions using preheat, aqueous-processing and postbaking.

High-speed UV photopolymer plates There is a family of double-coated photopolymer plates with very high exposure sensitivity in the UV and violet light (365-450 nm) spectrum that can be exposed in conventional vacuum frames or step-and-repeat machines, and with enough sensitivity for CTP using platesetters with 365 nm lasers or the new low-power violet laser diodes developed for use with DVD players. The family consists of four primary plate groups with 40%, 20%, 4% and 2% of the UV energy necessary to expose a *standard* UV sensitive conventional plate which takes about 250 millijoules of UV energy per square centimeter. The exposure is accomplished by light and heat polymerization which produces images with 1-99% dots in a 150-line screen image and can produce 300-line screen images in conventional and CTP imaging.

Waterless plates Waterless lithography is a planographic process like lithography but it prints without water and thus does not require an ink-water balance in printing. The process eliminates all of the disadvantages caused by the need for an ink-water balance, but retains all of lithography's advantages of reasonable plate costs, ease of makeready, high speed, high ink densities and print contrast with superior shadow detail, and exceptional print quality of fine screen images up to 500 lpi.

Waterless plates consist of ink on aluminum for the printing areas and a silicone rubber for the non-image areas. Silicone rubber has very low surface energy and thus has the property of not wanting to be wet by anything, especially ink. However, under the pressure and heat of printing, ordinary litho ink has a tendency to smear over the silicone and cause scumming or toning. Systems have been developed for waterless printing using special printing ink temperature control systems, and waterless plates in which the silicone rubber is cured by exposure. Inks have been developed that print with fixed temperature ranges. There are both positive and negative plates. Besides special inks and temperature controlled inking systems, waterless printing requires the use of good grades of paper to avoid accumulation of debris on the blanket caused by linting papers.

Bimetal plates These plates use presensitized polymer coatings. They consist of a metal base with one or more metals plated to it. There are two types of bimetal plates: (1) copper plated on stainless steel or aluminum and (2) chromium plated on copper. (The copper can be plated on a third metal which becomes the base.) Bimetal plates are the most rugged and also the most expensive of lithographic plates, but they are capable of runs in the millions.

Bimetal plates are easiest to run because they are almost indestructible. Should anything happen to the plate on press (the copper may refuse to take ink or the non-printing area may scum), a single acid treatment restores the plate to its original condition. With other types of plates, treatments used on press to restore ink-receptive areas are often injurious to water-receptive areas and vice versa. Another advantage of bimetal plates is that dot sizes can be controlled in the etching process. This is the main reason for their comeback especially for web off-set printing of publications and long run packaging. On diazo and photopolymer plates exposure is very critical as it controls both plate wear and dot gain. The amount of exposure required to obtain good plate wear often produces excessive or erratic dot gain in printing (dot gain on plates made from negatives and dot sharpening on plates made from positives).

On-press processed plates This is a new platemaking technology. The plates are exposed in a vacuum frame and mounted on the press without processing. The processing occurs on the press during makeready by about 20 revolutions in contact with the regular dampening solution which disperses the coating in the non-image areas, and about 10 revolutions of the ink rollers that absorb the loosened coating.

Computer-to-Plate (CTP) Systems

A CTP concept was tried in 1975 using electrostatic plates, but they did not have the resolution or the exposure speed for commercial printing. CTP plates using silver-halide coatings on a polyester base were introduced in 1982 and used for single color and loose register spot color printing. It was not until 1990 that lithographic plates on metal bases with the quality and speed necessary for four-color process commercial printing were introduced.

These were silver halide and high-speed dye-sensitized photopolymer plates. A hybrid silver halide mask coated on a metal base photopolymer plate was introduced in 1993. In 1995, hybrid plates with other types of masks and several thermal plates were introduced. The thermal plates used laser ablation, laser ablation transfer and thermal crosslinking technologies. A no-process ink-jet plate system is in limited use.

CTP has not advanced as rapidly as projected due mainly to the increased cost of CTP plates which range from 25% to 75% higher than high-speed UV photopolymer plates that can be used both as conventional and CTP plates.

High-speed UV photopolymer plates These plates are described on page 121. They are unique in that they can be used for both conventional printing and CTP which means that CTP can accomplish its objective of saving film and processing costs without increasing plate costs. There are almost a dozen platesetters that can use these plates. For printing plants using both conventional and CTP processes, they can save on equipment and consumable costs, as the same plates can be used for both processes.

Silver halide plates These high-speed plates with silver halide coatings have been available since 1982 on polyester bases for single- and spot-color printing. Coatings on 7 mil polyester are available for color printing. The first coatings on anodized aluminum for process color printing became available in 1990. Early coatings were color blind and were exposed by Argon Ion (488 nm) lasers. New coatings are panchromatic which can be exposed with red 670 nm lasers. The plates can also be exposed optically in vacuum frames or on step-and-repeat machines by high-intensity lamps. A main disadvantage of these systems is the processing. The processing solutions contain metal (silver) toxic pollutants that must either be transported to special chemical treating plants or be treated in-plant with silver recovery chemicals before draining into municipal sewers.

Dye-sensitized photopolymer plates Photopolymer coatings have high sensitivity to UV radiation. Like silver halide photographic emulsions, the light sensitivity of photopolymers can be extended into the visible light spectrum by the use of sensitizing dyes *(see page 64)*. The plates can be exposed by Argon Ion (488 nm) lasers or frequency doubled YAG (532 nm) green lasers. They are capable of run lengths over 250,000 without postbaking and over 1,000,000 impressions with baking. Most plates use aqueous processing.

Hybrid plates Hybrid plate technology uses two separate photo-sensitive coatings on metal plates. The bottom coating is a conventional photopolymer with good press experience and above it is a silver halide coating which can be exposed optically to films or digitally by lasers. During processing of the top coating, the bottom coating is exposed to UV light. After removal of the top coating and processing of the bottom one, printing is done from the photopolymer (bottom) coating. Serious deterrents to the use of this system are its use of two separate processors and the toxic wastes from the processing of the silver halide mask. Other masks have been tried including *thermal* which can be processed with the base coating, and *ink-jet* that requires no processing.

CTP thermal technologies Thermal technologies are being used because of the introduction of new low-cost high-power infrared (IR) laser diodes. These systems have three main advantages over visible light technologies: (1) *room light handling* thus eliminates special cassettes and darkroom loading; (2) *threshold binary imaging* which produces sharp, no-fringe high-contrast elements with exposures above the threshold temperature and no exposures below it — with little or no dot gain; and (3) *minimal* or *no processing*. Three thermal imaging technologies are in use: (1) *thermal laser ablation:* (2) *thermal laser ablation transfer* from donor sheets to acceptor plates; and (3) *thermally activated acid generated crosslinking*.

Thermal laser ablation plates Two types of these plates are in use: (1) *dry* (waterless) and (2) *wet* (ink/water balance). The dry waterless plates have a metal base coated with thin layers of polyester sputtered titanium and a silicone rubber. Wet plates are the same except they have a top layer of polyvinyl alcohol (PVA) in place of the silicone coating. The platesetter is equipped with special IR laser diodes with a wavelength of 860 nm that burns tiny holes as small as 5 microns (0.0002 inch or 0.005 mm) in the top and sputtered aluminum layers leaving slightly indented image elements. No chemical processing is required. Because of the no-liquid processing feature, most thermal laser ablation plates are imaged directly on-press (CTPs). The slightly dry debris from the laser ablation is wiped off the plates with a cloth before using. The plates are capable of runs up to 200,000 impressions.

Thermal laser ablation transfer plates These use a digital file to drive a laser at a special donor sheet that transfers spots of ink receptive polymer to a receiver plate in a platesetter. Spots as fine as 5 microns and runs up to 200,000 are claimed for the plates. Main problems with this type of system are the need for special modifications to the platesetter, cost of the plates and disposal of the donor sheets.

Thermal crosslinking technology In this type of system, exposure to an IR (830 nm) laser releases an acid in a special coating. The acid catalyzes the crosslinking of two resins in the coating. Early coatings required a pre-heat step prior to wet aqueous chemical processing. Post baking produces plates capable of over 2,000,000 impressions. The same crosslinking reaction occurs with UV radiation so the plates can also be exposed in a vacuum frame using a UV light source. Because of this dual sensitivity the plates must be handled in yellow light. These plates have very high resolution because of the binary image elements with essentially no dot gain. Continuing research has already eliminated the pre-heat step and may eventually eliminate all processing.

No process ink-jet technology The ideal plate for CTP is one that requires no processing or post treatment; can be produced in simpler and lower cost platesetters; and can be used immediately after imaging or after storage without additional treatment. *Ink-jet* imaging appears to have this potential. It has three appealing properties: (1) requires no processing; (2) does not use lasers or other optical imaging devices; and (3) platesetters for ink-jet are vertical with smaller footprints than laser systems and are lower in cost. Ink-jet systems have some speed and resolution limitations which are being improved.

Projection Plates
Some of the CTP plate systems can be used in the computerized projection imaging system described on page 118. The new projection systems for poster and book printing are designed to use the illumination level common in cameras, so plates with the speed of lithfilms are required. Ordinary diazo and photopolymer presensitized plates have exposure speeds 100 to 1,000 times slower than lithfilms. Four of the families of plates introduced for CTP platemaking have UV and/or visible light sensitivity so they can be used for projection platemaking. They are: (1) the high-speed UV photopolymer plates, (2) silver halide coatings on plastic and metal bases, (3) the new high-speed dye-sensitized photopolymers, and (4) the thermal crosslinking plates.

LETTERPRESS PLATES
Letterpress is a relief printing process using plates or imaging materials with the image areas raised above the non-image areas. As described in the section *Introduction to the Printing Processes, page 26,* letterpress has declined in use. When letterpress was in vogue, photoengravings made on 16-gauge (0.065″) zinc, magnesium or copper metal using powderless etching processes were used for direct printing or as molds for duplicate plates. The duplicate plates were *stereotypes* used mainly for newspaper printing; *electrotypes* used for long run magazine, catalog and commercial printing; and *plastic* and *rubber* plates used for specialty printing.

Wraparound Plates
Wraparound plates, as used for *dry offset (letterset)* which uses an offset lithographic press without dampening or printing on narrow web presses are made in one piece to be wrapped

around the plate cylinder of a press. All copy is in proper position for printing. Setup or makeready time is substantially reduced. Plates are plastic or metal, ranging from 0.017″ to 0.030″ in thickness so they can be bent to fit into the plate cylinder clamps.

Photopolymer Plates

Photopolymer letterpress plates are precoated and can be used as original (or direct) and wraparound plates. A nylon-based plate, on a steel base and mounted on magnetic cylinders, is the most popular plate used in letterpress for magazine, commercial and narrow web printing.

FLEXOGRAPHIC PLATES

Flexography is also a relief process and its plates are similar to letterpress plates except they are softer and resilient. Rubber plates are used extensively in flexography for printing envelopes, bags, tags, wrapping paper, corrugated boxes and milk cartons. The introduction of special photopolymer plates extended the use of flexography to many other printing markets such as newspapers, flexible films, pressure sensitive and other labels, and corrugated boxes and packaging.

CTP Technology for Flexography

Digital imaging of rubber flexographic plates and cylinders by laser engraving has been used since the late 1970s. Early systems suffered from slow speed and quality deficiencies due to vertical walls of image elements which caused excessive dot gain especially in highlight areas.

New systems using means of producing conical dot walls and the development of new laser systems with higher switching speed and laser engravable plates with rubber/polymeric coatings have produced laser engraved images with halftone dots ranging from 2% to 98% on 120 lpi process color prints. The newer engraving systems produce flexographic cylinders with seamless joints, and cylinders and plates with step-and-repeat, staggered and nested images for packaging printing.

GRAVURE PLATE AND CYLINDERMAKING

Gravure is an intaglio process in which the image areas are cells in a thin copper shell plated on a steel cylinder, and a steel doctor blade keeps the non-printing areas clean during printing. Gravure differs from other processes in that it is capable of printing varying amounts and densities of ink to produce images that simulate continuous-tone images. The intaglio cells that compose the images vary in volume corresponding to the tonal values in the images. Two processes for producing the cells represent two different types of cells. They are: (1) conventional gravure with constant area variable depth cells and (2) electromechanical engraving (EME) that produces variable area–variable depth cells.

TWO TYPES OF GRAVURE

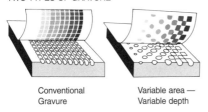

Conventional
Gravure

Variable area —
Variable depth

Conventional gravure uses a carbon tissue, on which continuous tone positives are exposed and an overall square dot pattern is made by a special gravure screen with a 3:1 ratio of black spaces to clear lines. The exposed tissue is attached in position on the gravure cylinder, and etching of the cells is done chemically or with powderless etching similar to that used for copper photoengravings. The purpose of the overall screen is to create walls for the cells and provide a common surface for the doctor blade to ride over. Although the images are of high quality, conventional gravure is not practical for runs over 50,000 because the highlights are so shallow they gradually wear and disappear due to abrasion by the doctor blade and paper.

Electromechanical engraving Most gravure cylinders are made by electromechanical engraving systems consisting of three parts: (1) an input unit that is a rotating drum on which photographic prints are mounted in position and scanned by

ELECTROMECHANICAL ENGRAVING

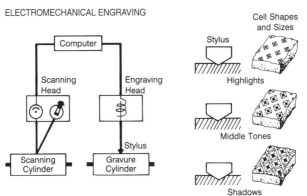

Computer

Scanning
Head

Engraving
Head

Stylus

Scanning
Cylinder

Gravure
Cylinder

Cell Shapes
and Sizes

Stylus

Highlights

Middle Tones

Shadows

one or more reading heads; (2) a system computer which processes the image densities recorded by the reading heads and converts them to electrical impulses that are (3) transmitted to an output unit. This consists of a rotating copper-plated cylinder on which are mounted one or more engraving heads with diamond styli that engrave gravure cells in the cylinder corresponding to the strength of the analog signals. These cells are produced at 4,000–6,000 per second.

New laser gravure process A new direct digital laser etching process has been introduced for gravure which uses a special alloy that can be plated on steel like copper but has much higher laser etching efficiency than copper which reflects laser beams. It can use the same digital files as the EME but is capable of etching up to 70,000 175 lpi cells per second (about 10-15 times the cell-etching capacity of EME machines). This new process will simplify the imaging of gravure cylinders and eventually reduce their cost.

Photopolymer gravure Photopolymer plate systems for gravure have been developed. They consist of photopolymer coatings on stainless steel plates which can be mounted on magnetic cylinders. These plates are similar to photopolymer relief plates except that the images are depressed instead of in relief, and a steel or special plastic doctor blade is used. These plates make gravure a viable process for runs below 100,000 and can make gravure competitive with lithography and flexography in this run category, especially in packaging.

Computer-to-cylinder systems for gravure Gravure was the first printing process to use computer-to image carrier technology. In 1980, interfaces were introduced between color electronic prepress systems (CEPS) and electromechanical engravers (EME). Instead of using prints on the EME, the digital information from the CEPS is fed directly to the EME computer which converts the digital signals to electrical impulses that are fed to the diamond styli in the engraving heads that produce the printing cells. This process is also called *Filmless Gravure*. The digital information can be used to drive the new laser gravure etching system and produce cylinders at a faster rate.

SCREEN PRINTING

Versatility is the principal advantage of screen printing. Any surface can be printed (wood, glass, metal, plastic, fabric, cork, etc.) in any shape or design, any thickness and any size. In advertising, screen printing is used for banners, decals, posters, billboards, car cards, counter displays, menu covers, etc.

There are many methods of making screens for screen printing. The screen is a porous material and the printed image is produced by blocking holes or *pores* of the screen representing the non-printing areas. Early screens were made manually by painting the image on a silk screen mounted on a wooden frame. Masking materials were used to block out unwanted areas. Today, both hand-cut stencils and photomechanical means are used. In the photomechanical method, the screen is coated with a light-sensitive emulsion, exposed to a halftone positive film placed in contact with the screen, and the coating in the unexposed areas is developed to form the image through which the ink is forced on to the substrate using a squeegee. The exposed areas form a hardened stencil which prevents ink penetration in these areas.

Rotary screens are made with the same techniques as flat screens using overall perforated cylinders. They are also made by plating a metal cylinder electrolytically on a steel cylinder, removing the metal cylinder after plating, applying a photosensitive coating to the cylinder, exposing it through a positive halftone image and etching out the image areas to form pores in the cylinder. On rotary screen presses, the ink is pumped into the cylinder and the squeegee which is inside the cylinder controls the flow of ink to the substrate.

PLATES FOR DUPLICATING

Some unconventional plates are used in duplicating (reprography) that have lithfilm speeds, but these are generally short run, with some quality limitations. They are *photographic, diffusion transfer, electrostatic* and *direct image* plates.

Photographic plates using chemically-tanned silver halide emulsions are used for duplicating printing. One is a *camera plate* system in which the plate is on a plastic base and is made directly in a camera, processed and mounted on the press. Another is a paper base coated with two photographic emulsions and a third coating containing photographic developer. During exposure and processing the gelatin in the unexposed or image areas of the copy are tanned and become ink-receptive, while the exposed non-image areas remain water-receptive. A third type is similar to the high-speed computer-to silver halide plates described on page 122 which can be made from digital data.

Diffusion transfer plates produce a negative of the image to be reproduced on an intermediate surface, and the final positive image is transferred by diffusion to a receiver material. Some quality is sacrificed in the transfer, but plates are simple to make and some are capable of runs up to 50,000.

Electrophotographic plates are used mainly in duplicating on offset duplicators and copier/duplicators *(see page 30)*. They do not have the necessary resolution for commercial printing.

Direct image plates can be prepared by typing, drawing or lettering directly onto a paper or thin metal master using special ribbons, pencils, crayons and inks. The old lithographic stones are examples of direct image plates. Artists still draw on stone.

CTP Technology for Duplicators

There are several manufacturers of digital duplicators. The more advanced are the porous type. Computer interfaces have been developed that use scanners or a PostScript-based system that sends digital signals to a thermal head which creates the perforations in the stencil. Pages can also be composed using a digital layout tablet and a light pen.

Printing

As described in the section *Introduction to the Printing Processes (page 24)* there are two types of printing processes in general use: (1) plate, impact or pressure printing, and (2) plateless, non-impact or pressureless printing. Plate systems are the major printing processes: offset lithography, letterpress, flexography, gravure and screen printing which are used to print quantities of reproductions for the newspaper, commercial, magazine, catalog, book, packaging, label and other large printing markets. Plateless printing processes are the digital printing systems: electrophotographic presses, electrostatic copiers and printers, ion or electron deposition, magnetographic, ink-jet, thermal transfer and dye sublimation printing systems used mainly for short-run, variable information, or on-demand printing. This section is divided into two parts: (1) plate printing processes and (2) digital printing processes.

Each printing process has distinguishing quality and appearance characteristics. Differences between the final printed images are not as noticeable as they were years ago. Offset lithography, letterpress and gravure are all capable of reproducing the same art and copy with equally satisfactory results, and flexography is rapidly approaching their quality. There are still some distinguishing characteristics, but these are minor. In general the major plate processes (lithography, letterpress, UV flexography and gravure) have higher resolution and print quality than the plateless printing processes.

Letterpress, and to some extent flexography, are distinguished by sharp, crisp printing, grainy images with sharp breaks in gradient tints and vignettes, and slight embossing of type on the back of printed sheets. Offset lithography is characterized by soft, smooth transitions of color and tones and slight differences in color balance throughout a run. Waterless printing has very high resolution, ink density and print contrast. Gravure has a long tone scale with strong, saturated colors and slightly ragged type. In most cases, however, factors such as economics, availability of equipment, length of run and speed of delivery are the main considerations in selecting a printing process rather than the appearance of the image.

PRINTING PRESSES

The production machines of the major printing processes are the *presses*. In general, a printing press must provide for: secure and precise mounting of the image carrier (and, in offset lithography, a blanket); means for feeding paper for printing and delivering printed paper; accurate positioning of the paper during printing; conveying the paper through the printing units to the delivery; storing and applying ink (and, in lithography, a dampening solution) to the plate; and accurately setting printing pressures for transfer of the inked image to the paper.

Presses are either sheetfed or roll- (web-) fed. Much commercial printing is printed on sheetfed presses. Magazines, newspapers, mail-order catalogs, books and other long-run work are printed on web-fed presses. Presses may be single color or multicolor. Usually each color on a multicolor press requires a separate complete printing unit of inking, plate and impression mechanisms. A two-color press would have two such units, a four-color press would have four, etc. Some presses share a common impression mechanism among two or more printing units. Packaging and other special purpose presses may have combinations of offset lithographic, letterpress, flexographic and gravure units. A *perfecting* press is one that prints both sides of the paper in one pass through the press. Most web and many sheetfed presses are designed to perfect.

OFFSET LITHOGRAPHY

While *letterpress* was the first printing process, *offset lithography* has replaced it as the leading printing process. Lithography is responsible for almost half the printing using printing plates. It can be printed *direct* but over 99% of lithographic printing uses the *offset principle*. In fact the word *offset* has become synonomous with lithography. The offset press is responsible for four important advantages of lithography: (1) the rubber blanket surface conforms to irregular printing surfaces, resulting in the need for less pressure and makeready, and improved print quality of text and halftones on rough surfaced papers; (2) paper does not contact the printing plate, increasing plate life and reducing abrasive wear; (3) the image on the plate is right reading rather than reverse reading; (4) less ink is required for equal coverage, drying is speeded, and smudging and set-off are reduced.

Offset Presses

Offset presses have three printing cylinders (plate, blanket and impression) as well as inking and dampening systems. On most offset presses, as the plate cylinder rotates, the plate comes in contact with the *dampening* rollers first, and then the *inking* rollers. The dampeners wet the plate so the non-printing areas repel ink. The inked image is then transferred to the rubber blanket, and paper or other substrate is printed as it passes between the blanket and impression cylinders.

Makeready, which is the work done to set up a press for printing, is minimal as compared with letterpress. The wraparound

PRINCIPLE OF OFFSET LITHOGRAPHY

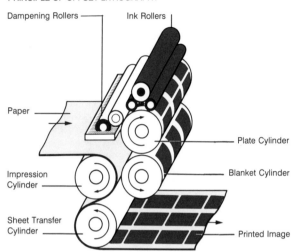

Dampening Rollers

Ink Rollers

Paper

Plate Cylinder

Impression Cylinder

Blanket Cylinder

Sheet Transfer Cylinder

Printed Image

plates can be moved slightly for proper register. The resilient rubber blanket compensates for the variable thickness and texture of paper stocks, largely eliminating a source of considerable trouble in other printing processes. A wide range of papers or other substrates can be used. Halftones can be printed with text and solids on both rough and smooth surface papers. The most time-consuming operation is setting the ink-water balance.

Sheetfed presses Most commercial sheetfed printing is on 25″ x 38″ to 38″ x 50″ presses. They can print at speeds up to 15,000 impressions per hour and are made as single or multicolor presses up to 10 units. Larger presses, up to 64″ wide, are used mainly for printing specialties like greeting cards, gift wrappings, posters, labels and packaging. Sheetfed printing has the advantages that (1) a large number of sheet or format sizes can be printed on the same press, and (2) waste sheets can be used during makeready, so good paper is not spoiled while getting position, color or ink-water balance up for running.

Sheetfed presses are available with perfecting features. Some are single color with blanket-to-blanket design. Many newer presses with two or more units have convertible perfecting units. Most perfectors have fixed configurations.

Newer multi-unit sheetfed presses are available with coating towers and/or extended deliveries to assist ink setting and provide space for infrared drying or UV ink-curing units.

Web offset is used to produce a wide variety of long-run commercial printing. The 38″ 16-page web offset press has almost completely displaced large sheetfed presses 60″ and larger for books, periodicals and commercial printing. The half-size, or 8-page web (17″–31″ width), has made inroads on the 25″ x 38″ to 38″ x 50″ sheetfed market. The 32-page presses make web offset competitive with gravure in very long-run printing.

Speed is the main advantage of web offset. Some presses are rated at speeds of 3,000 feet per minute. Most web offset presses have inline folders where various combinations of folds convert the web into folded signatures. Some presses have variable folders to handle different types of printed products. Other inline operations include paste binding, perforating, numbering, rotary sheeting and slitting.

A disadvantage of most web offset (and web letterpress) is that it has a fixed cutoff (or print length). A major advantage of rotogravure and flexography is that cylinders with different diameters can be interchanged on the press allowing for different cutoffs. Web offset presses are available with interchangeable printing units and variable cutoffs.

There are four types of web offset presses:

(1) The **blanket-to-blanket press** has no impression cylinders. The blanket cylinder of one unit acts as the impression cylinder for the other, and vice versa. Each printing unit has two plate and two blanket cylinders. The paper is printed on *both* sides at the same time as it passes between the two blanket cylinders. In all web printing the grain of the paper is in the direction of travel of the paper through the press. The grain of the paper is around the circumference of the cylinder. This is opposite to the grain direction of the paper in most sheetfed printing which is parallel to the axis of the cylinder. The latest developments in web offset presses are gapless plate and blanket cylinders. These eliminate the vibrations caused by the cylinder gaps, and make printing speeds of 3,000 fpm possible.

(2) The **blanket-to-steel press** has printing units similar to a sheetfed offset press, except that the plate and blanket cylinder gaps are very narrow. Each unit prints one color on one side; additional units are required for additional colors. To print the reverse side, the web is turned over between printing units by means of *turning bars*. This type of press is used extensively for printing business forms, computer letters and direct mail advertising.

(3) **Variable-size presses** Gravure and flexography have variable-size cutoff, or print-length capability as different diam-

BLANKET-TO-BLANKET PRINTING UNIT

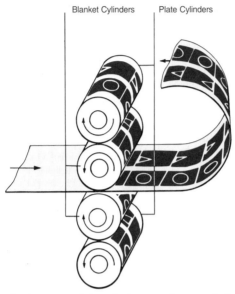

Blanket Cylinders   Plate Cylinders

eter image or plate cylinders can be mounted on the press. Variable print-length presses for lithography have been available since the 1960s. These presses use removable printing units, inserts or cassettes for one-side and blanket-to-blanket two-side printing, in widths up to 48″ and circumferences up to 40.5″.

(4) **Keyless offset** uses keyless inking systems. Some systems are similar to the flexographic systems using anilox rolls and reverse angle doctor blades. Keyless offset is based on the concept of using fresh ink for each revolution by removing residual inks on the inking drum after each revolution. Press manufacturers in Japan, Germany and the U.S. are making keyless offset presses, primarily for newspaper printing. They use soft or liquid inks as opposed to stiff or thixotropic inks used for commercial lithographic printing.

Small offset press is another category of offset presses in size 14″ x 20″ with multi-units and the print quality of larger presses, used for spot and process color printing. These presses produce higher quality printing than offset duplicators and digital presses. They are used mainly for high quality short runs as high as 5,000 or more sheets.

Offset duplicators are small offset lithographic presses which are used for fast, good quality reproduction of copies in sizes from 3″x 5″ up to 12″x18″. They are ideal for low-cost printing of business forms, letterheads, labels, bulletins, postcards, envelopes, folders, reports and sales literature.

Machines are made for simplified operation and convenience. The offset duplicator is a compact, reliable, high-production machine with many built-in features for fast job changeovers and minimum makeready. It can print on sheet stock from lightweight onionskin up to cardboard at speeds of up to 12,000 impressions per hour (iph). Web-fed duplicators are more limited in use, but are capable of speeds of up to 25,000 iph.

Dampening systems on offset presses transfer the dampening solution to the plate. In the direct-feed continuous integrated type of dampening system, the fountain solution containing alcohol or an alcohol substitute is metered to the plate through the inking system, or can be applied directly to the plate as in other systems. In general, this type of dampening system uses less water and reduces makeready time and paper waste at start-up of the press. Because of the cost of isopropyl alcohol and potential health hazards in its use, a number of new fountain solutions have been developed to reduce or replace the alcohol in this type of dampening system.

Fountain solutions have traditionally used pH or acidity as a means of control which worked well when acid solutions were used. Problems were encountered when buffered and alkaline solutions were introduced. These problems have been essentially solved by measuring both the pH and conductivity of the solutions. Also, special fountain solutions have been developed with oxidizing chemicals that accelerate the setting and drying of inks and reduce the need for anti-set-off sprays.

Inking systems transport ink from the ink fountain to the plate. All systems use composition rollers. Some have plastic-coated rollers and others have copper-plated steel rollers to prevent stripping of ink on the distributors. Many inking systems, especially on web presses and sheetfed presses for printing waterless plates, are water-cooled for temperature control. The ink form rollers in contact with the plate have different diameters and hardness to assure good ink transfer and prevent roller streaks in the printing.

Inline sheeters and on-press coating & finishing Some sheetfed presses, especially for packaging, use inline sheeters at the feeder to eliminate the difference in cost between sheet and roll papers and boards.

Most new sheetfed presses can be equipped with *coating units* or *towers*, with extended deliveries and *infrared* or *UV curing* units to eliminate or reduce the need for anti-set-off sprays and off-press coating or varnishing. Other inline finishing functions on presses include imprinting, perforating, numbering, stitching, paper gluing, applying remoistenable glue, making additional folds in the web direction, hole punching, rotary or flatbed die-cutting, web rotary cutter-creaser, etc.

Temperature controlled presses are improving the consistency, predictability and productivity of the printing process. Temperature control for cooling ink rollers is being used on sheetfed (waterless printing) as well as web presses. Heating the side frames of web offset presses has stabilized printing on the press by reducing the effects of heat buildup of cylinder bearings, journals and bearers. Advantages gained include lower dot gain, less fanout, less water consumption, better color consistency, less bearing, bearer and gear wear, lower energy costs and higher productivity.

Waterless printing is like lithography but without dampening. Silicone rubber is used in the non-image areas on the plates instead of water to repel inks. Specially formulated inks and temperature controlled inking systems are required to assure constant ink viscosity and prevent toning of the non-image areas. Both sheetfed and web-fed presses are used for waterless printing by disengaging the dampening systems.

The elimination of the ink-water balance has provided waterless printing with many advantages including faster make-readies, better color consistency, less dot gain, higher print resolution (up to 500 lpi screens), higher ink densities and print contrast, more saturated colors and shadow detail, less paper waste and higher productivity than conventional lithography.

Screenless Printing

One of the oldest photomechanical printing processes, *collotype*, reproduces illustrations in continuous tone or without halftone dots. Until recently, it was the only screenless printing process. Screenless printing can also be done by lithography and can be simulated by the use of frequency modulated (FM) screened images.

Collotype is a reproduction process which uses bichromated gelatin as a printing medium and is capable of high-quality reproduction in runs from 100 to 5,000. Platemaking and printing are extremely critical compared with the other printing processes. It has been used for posters and transparencies printed on both sides for back-lighted displays. Collotype has also been used for fine art reproductions, mounted displays and counter cards.

In screenless printing by lithography, continuous-tone films, very fine screen images (300–600 lpi), or frequency modulated (FM) digital images, are used on special plates. These processes have practically replaced collotype in the U.S. The main advantages of screenless printing are high-resolution, image sharpness, no screen moiré and greater purity of color especially in tints and middle tones.

LETTERPRESS

Letterpress, the first printing process, started with a converted wine-press. Through the years three basic types of presses were developed and used for letterpress printing: *platen, flatbed cylinder* and *rotary*. In addition there is a *belt press* for continuous inline printing and finishing.

Platen presses carry both the paper and the type form on flat surfaces known as the *platen* and the *bed* which open and close like a clamshell. The press is still used for short-run job printing such as announcements, invitations, name cards, stationery, etc. Larger platen presses are used for embossing, diecutting and scoring.

Flatbed cylinder two-revolution presses with a moving flatbed that held the form and a fixed rotating impression cylinder are now obsolete. Single revolution vertical presses where the bed is in a vertical position and both form and cylinder move up and down in a reciprocating motion are still in use.

Rotary sheet-fed presses with two cylinders are the fastest and most efficient of the three types, but, due to the declining market for letterpress, this press, like the flatbed cylinder press, is no longer manufactured in the U.S.

Web-fed rotary presses print a continuous roll or web of paper on both sides, one at a time, as it passes through the press. Most letterpress multicolor web-fed rotaries have been of the common impression cylinder type that used special heat-set inks and dryers. These presses have been used for all types of printing, from newspapers to color work in magazines and catalogs, but their use has declined.

Narrow web rotary presses (under 24″ width) using photopolymer plates and UV inks have revitalized the use of letterpress in the label and other printing markets. The presses have as many as 12 printing units and use other printing processes

PRINCIPLE OF
ROTARY LETTERPRESS

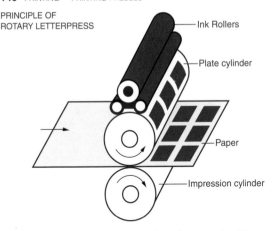

Ink Rollers

Plate cylinder

Paper

Impression cylinder

in-line as well as letterpress, such as flexography, lithography, gravure, screen printing and die-stamping.

Belt press is an automated inline press which prints, collates and binds a complete book in one pass through the press. It has two belts on which the plates for all the pages in the book are mounted. In printing, a roll of paper is fed in contact with the first belt which prints all the pages for one side, after which the ink is dried, the paper turned over, run in contact with the second belt, dried, slit into ribbons, folded, cut into signatures, collated into books and adhesive-bound into preprinted covers.

Makeready
The major problem causing the decline in letterpress printing is long makereadies. This is due to the variable pressure exerted by different size image elements in printing. The same amount of pressure, or *squeeze*, needed for ink transfer exerts greater pressure per unit area on small highlight dots than on larger shadow dots. Expensive manual makeready is needed to even out the impression so that highlights print correctly and do not puncture the paper and the solids print with even density. Precision electros, wraparound plates and pre-makeready systems helped to reduce makeready cost but not enough to save letterpress. Photopolymer plates and UV inks on narrow web presses are helping letterpress make a comeback. In flexography, makeready is not critical because the resilient plates distort and compress to balance the pressure differential. The distortion, however, can limit the quality of the screened image and affect the register in printing.

FLEXOGRAPHY

Flexographic presses are web-fed machines of three types: (1) *stack type* in which two or three printing units are placed vertically in stacks *(see illustration on page 27)*. A press may consist of two or three stacks, with unwind, rewind, sheeter or cutter and creaser; (2) *central impression cylinder* which is like the common impression rotary letterpress and is used extensively for printing flexible films; and (3) *inline* which is similar to a unit type rotary press.

FLEXOGRAPHIC CENTRAL IMPRESSION CYLINDER FOUR-COLOR PRESS

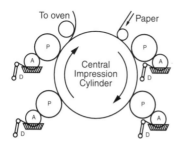

P - Plate cylinder
A - Anilox roller
D - Reverse angle
Doctor blade

Flexography is the least expensive and simplest of the printing processes used for decorating and packaging printing. It uses rubber or special photopolymer plates and water- or solvent-based inks in simple two-roller *anilox* inking systems. Good quality printing is produced on paper and flexible films using photopolymer plates, reverse-angle doctor blades and ceramic anilox rollers in central impression cylinder presses. These developments have qualified flexography for the book, newspaper insert, publication and packaging printing markets.

In printing on *narrow web presses* the use of special *UV inks* has been responsible for significant improvements in print quality, such as the printing of 3% dots in 150 lpi screen images. With UV inks, flexography is approaching the quality of lithographic printing in many printing markets.

GRAVURE

Rotogravure printing units consist of a printing cylinder, an impression cylinder and an inking system. Ink is applied to the printing cylinder by an ink roll or spray, and the excess is removed by a doctor blade and returned to the ink fountain. The impression cylinder is covered with a rubber composition that presses the paper into contact with the ink in the tiny cells of the printing surface. Gravure inks use volatile petroleum solvents and dry almost instantly. Hot air dryers are used between printing units to speed up drying. Therefore, in color printing, ink trapping is not a problem since each succeeding color is printed on a *dry color*, rather than on one which is still wet as in letterpress and offset. For color printing, presses use electronic systems for automatic register control. Printing cylinders are chromium-plated for press runs of one million or more. When the chromium starts to wear, it is stripped off and the cylinder rechromed.

Gravure is used in packaging for quality color printing on transparent and flexible films (any cut-off length is possible by changing the printing cylinder), and for printing cartons, including die-cutting and embossing which can be done inline on the press. Among the specialties printed by gravure are vinyl floor coverings, upholstery and other textile materials, pressure-sensitive wall coverings, plastic laminates, imitation wood grains, tax and postage stamps and long-run heat transfer patterns. Most very long-run magazines and mail order catalogs are printed by gravure. This market is decreasing in run lengths in the U.S. as demographic editorial and advertising editions and target marketing increase.

Large publication gravure presses are almost completely automated and can print webs up to 120″ (3 meters) wide. For competing with web offset in the shorter run market, the *pony press* used has the *double ending* principle and prints 40″ wide signatures. For printing floor coverings, multicolor gravure presses can

PRINCIPLE OF GRAVURE

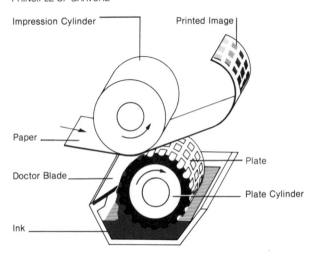

Impression Cylinder

Printed Image

Paper

Doctor Blade

Ink

Plate

Plate Cylinder

print webs up to 150″ wide. Presses used for packaging materials usually have webs from 40″ to 60″ wide with up to eight printing units. While speeds of up to 3,000 feet per minute are used in publication and catalog printing where runs are in the millions, average production speeds of up to 800 feet per minute are more realistic on other types of work with shorter runs.

Electrostatic assist An image quality defect in gravure is *dot skips* in highlight areas. It is due mainly to paper surface defects. The defect is correctable with *electrostatic assist*. An electrostatic charge on a conductive blanket on the impression cylinder is used to increase the transfer of ink from the gravure cells to the paper. The electrostatic charge increases the ink transfer by changing the shape of the meniscus of the ink in the cells allowing low spots in the paper that do not contact the cylinder to be wet by the ink. This increased ink transfer makes it possible to print with less pressure between the image and impression cylinders.

Sheetfed gravure presses operate on the same rotary principle as rotogravure. The preparatory work is identical. The image is etched on a cylinder or on a flat flexible sheet of copper which is then clamped around the plate cylinder of the press. Sheetfed gravure is used primarily for short runs and press proofing. Because of the high quality, it is used for art and photographic reproductions and prestige printing such as annual reports.

Offset gravure has been used for printing wood grains and in packaging. A converted flexographic press is used. The anilox roller is replaced by a gravure image cylinder and doctor blade for printing the image, and the plate cylinder of the flexographic press is covered with a solid rubber plate or a blanket.

STEEL-DIE ENGRAVING

Steel-die engraving is also an intaglio process in which the die is hand or machine cut, or chemically etched to hold ink. The plate is inked so that all sub-surfaces are filled. The surface is wiped clean; the paper is slightly moistened and forced against the plate with tremendous pressure, drawing the ink from the depressed areas. This produces the characteristic embossed surface, with a slightly indented impression on the back of the paper. Copper plates are used for short runs of one-time use (invitations and announcements). For longer or repeat runs such as letterheads, envelopes, greeting cards, stamps, money and stock certificates, chromium-plated copper or steel plates are used in a die-stamping press.

SCREEN PRINTING

Some screen printing is done by hand with very simple equipment consisting of a table, screen frame and squeegee. Most commercial screen printing, however, is done on power-operated presses of two types. One uses flat screens which require an intermittent motion as each screen is printed. The other uses rotary cylindrical screens with the squeegee mounted inside the cylinder, the ink pumped in automatically, and magnets used opposite the screens to control squeegee pressure. The rotary presses are continuous running, fast and print continuous patterns with little difficulty.

SCHEMATIC OF A FOUR-COLOR ROTARY SCREEN PRESS

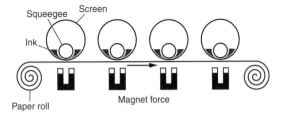

The amount of ink applied by screen printing is far greater than in letterpress, lithography or gravure which accounts for some of the unusual effects in screen printing. Because of the heavy ink film, the sheets must be racked separately until dry or passed through a heated tunnel or drier before they can be stacked safely without smudging or set-off. UV curing ink has simplified drying and is helping to promote greater use of screen printing, especially for coating and labels.

Screen printing is a unique short-run process that prints on almost any surface, and both line and halftone work can be printed. It is used for art prints, posters, decalcomania transfers, greeting cards, labels, menus, program covers, wallpaper and textiles such as tablecloths, shower curtains and draperies. It is particularly adapted to the printing of leather, metal, glass, wood, ceramic materials and plastics, in both flat and finished molded form. By printing an adhesive size and then dusting with cotton, silk or rayon flock, the finished design can be made to appear like felt or suede leather. It is also used for printing integrated circuits for electronics.

THERMOGRAPHY (Raised Printing)

Thermography is a process which creates special embossed effects in printing stationery, invitations, greeting cards and paper decoration, simulating die engraving without using costly engraving dies. Special non-drying inks are used in printing, either by letterpress or offset, and the wet inks are dusted with a powdered compound. After the excess powder on the non-printing areas is removed by suction, the sheet passes under a heater which fuses the ink and powdered compound. The printing swells or *raises* in relief to produce an engraved effect.

HEAT TRANSFER PRINTING

This process is used for producing printed products in which images are printed on paper with special inks and are then transferred to the desired substrate by heat and pressure. The process was developed for producing images and/or patterns on polyester fabrics. The inks contain special sublimable dyes which transfer to the polyester materials under the heat and pressure used. The process is an economical way of preparing images for use on expensive materials such as polyester fabrics. The printing processes used most extensively for heat transfer printing are gravure for long runs, flexography for shorter runs, screen printing for very short runs and offset lithography for sizable runs on sheet or piece goods.

DIGITALLY IMAGED (DI) PRESSES

Conventional/Digital and Digital Printing Systems are described briefly in the section *Introduction to the Printing Processes (page 32)*. Also described are the Computer-to *(page 31)* Technologies; i.e., Computer-to-Plate (CTPe) and Computer-to-Plate-on-Press (CTPs) which are used in the Conventional/Digital processes, and Computer-to-Print (CTPt) which uses digital presses.

Digitally imaged, or direct imaged (DI) presses are offset lithographic presses using *thermal laser ablation* or other non-processing plates described on page 124. Because of the no-liquid processing feature, the plates can be imaged directly on-press using special press-mounted imaging units that contain the thermal lasers driven by digital files. This feature of on-press imaging (CTPs) shortens makereadies, assures tight register ($\pm\frac{1}{2}$ row of dots), and automatically sets ink fountain keys on the press as the plates are being imaged.

A spark discharge plate system was introduced in 1991 which was replaced by a thermal laser ablation system in 1993. From then until 1999, it was the only system used commercially with on-press imaging offered by two press manufacturers. Two similar systems using the same plates were announced, one in 1997, and the other in 1999. A fourth system using different plates and a dampening system started shipping in the year

2000. Two other systems using the thermal laser ablation water-less plates, one 22″ x 29″ with a six-over-six page press design, are in development.

SCHEMATIC OF A FOUR-COLOR SHEETFED DI PRESS

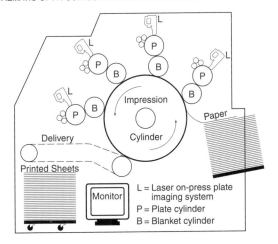

Other computer-to-plate-on-press platemaking systems in development use gapless sleeves, on-press coating, imaging and printing, followed by erasing and repeating the operation. As new high-speed no-process plate systems are developed, among the first applications that will be tried will be on-press imaging for CTPs use. This is expected to be a large growth area in digital printing as more press manufacturers license the system already in use or develop new systems using similar technology. It has been predicted that by the year 2005 most new presses will be of the DI design.

DIGITAL PRESSES

Digital printing uses digital presses which are the printing engines of the plateless processes. These are also the Computer-to-Print (CTPt) processes. This section describes the CTPt technologies that encompass the whole array of plateless printing technologies. These include: electrophotography (EP), ink-jet, ion or electron charge deposition, magnetography, thermal transfer, thermal dye sublimation and electro-coagulation. These processes are described briefly in the section on *Introduction to the Printing Processes (page 24)*. As pointed out they have some unique features conventional plate printing processes lack, such as:

- Most equipment is copier-like that can be handled in office environments.
- They are capable of variable printing from impression-to-impression.
- They require less manual skills than printing on conventional plate presses.

Very few, however, have achieved the speed and quality of most commercial plate processes, and, for printing quantities of the same reproductions, unit costs limit them to short runs of under 2,000 copies. The major markets that use these technologies are short run color, variable information, on-demand and distributed printing. These technologies are creating a new concept of printing. Instead of *print and distribute* which characterizes current conventional plate printing workflows, the use of these technologies can change the workflow to *distribute digital files* and *print on location*.

ELECTROPHOTOGRAPHY

Electrophotography (EP), also known as xerography, is the most widely used of the plateless printing technologies. Its principles are described on page 30. Developed by Xerox in the 1950s for copying, electrophotography was first used as a digital printing system in 1978 by what were called intelligent copiers, but were actually the first laser printers.

Electrophotography consists of a photoconductor which is charged in darkness with a corona discharge and exposed to an image by light that discharges the photoconductor where the light strikes it. The remaining charges are imaged with an oppositely charged dry or liquid toner. The image is transferred to a substrate and fused or fixed by heat, solvent vapor or other fixing method.

EP copiers use ordinary incandescent, fluorescent or strobe flash light sources to expose images to be copied. EP printing engines, generally known as *laser printers* use lasers or LEDs (light emitting diodes) as light sources driven by digital data from computer systems to produce the images to be printed.

Digital Presses

Digital printing became recognized as a strong contender in the short-run, variable information and on-demand color printing markets in 1993 with the introduction of two electrophotographic printing presses. One is a converted sheetfed single color small offset press with facilities for printing up to 7-colors on one or both sides of the paper. The other is a web design with facilities for printing up to 4-colors on both sides of the web. The sheetfed press has the following features:

- An organic photoconductor which can be used for many exposures is mounted on the plate cylinder.
- The inking system is removed and replaced with a corona charger for charging the photoconductor, a laser-imaging system for exposing the photoconductor, and toning nozzles for applying the liquid toner immediately after laser exposure of the corona-charged photoconductor.
- A heated blanket on the blanket cylinder.
- System for feeding printing paper on the impression cylinder, having it rotate on the cylinder one revolution for each color and turning it over for duplex printing before moving to the delivery.

SCHEMATIC OF A SIX-COLOR SHEETFED DIGITAL PRESS

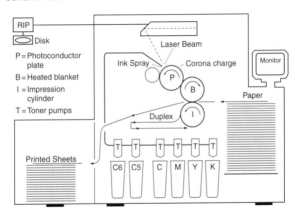

There are toner cartridges feeding toner nozzles corresponding to the number of colors in three modifications of the press-system — 4-color, 6-color, and 7-color. A special *Electro Ink* formulation uses micron-size pigment toners dispersed in a thermoplastic resin and diluted in a light mineral oil *(isopar)*. Speed of four-color printing (one side) is up to 2,000 sheets per hour for the 7-color press.

A variation of this system modifies the printing engine by overprinting the four to six images on the heated blanket, and making a single transfer of the combined images to the substrate. This modification has improved register and made possible the conversion to web printing. This system is used for short-run color package printing.

The web digital printing system has the following features:

- Eight photoconductor covered cylinders for four-color images on both sides of the paper.
- Light emitting diodes (LED) are used for exposing the eight image cylinders.
- Special dry small-particle dual component toners are used which are fused by heat after imaging.
- Speed is 1,050 sheets per hour — 70 two-sided pages per minute.

A schematic of this press is on page 33.

Digital Color Copier Printing Systems

When a color server or controller is added to an EP color copier the device can be used as a digital color printer for short-run color printing (1–2,000). The first color copier used for digital color printing was shown in 1990 using a PostScript controller. There are thousands of color copier printing systems now in use worldwide. Three main types of systems are:

- One type uses new 4-drum imaging engines that can print from 31 to 40 color pages per minute (one side) with 400 dpi resolution and 6 to 8 bits per pixel color print quality. These systems with color servers cost less than the presses.
- A second system uses clusters of two, four or six unit configurations of single color and 4-color EP engines driven by a special server, RIP and OPI software. A unique feature of this system is its job parsing software that allows each unit to operate as a single output device. When the job is completed, the printed stacks for each unit can be combined to produce a properly collated product. Cost of a 4-unit system is about a third the cost of the 4-drum systems.

- The third type uses modified single drum EP color copiers equipped with color servers. These have resolutions of 400 to 600 dpi and print 3–7.5 pages per minute (ppm). As of the year 2000, manufacturers produced over 20 types of color copiers equipped with color servers to produce color printers.

Digital Laser Printer Color Printing Systems

As already mentioned laser printers date back to 1978. Early systems suffered from low resolution (300 dpi and lower). In 1989, a 600 dpi single color laser printer was introduced that is capable of printing 135 documents up to 11"x17" size per minute. The first color laser printer was introduced in 1995, and since then a number of OEM laser printers are being used by over ten manufacturers to produce digital color printing systems with resolutions as high as 1,200 dpi and speeds up to 60 ppm (4-color) and 200 ppm (monochrome). Most of these color laser printers sell for less than $10,000.

Limitations of EP Digital Printing Systems

Users of EP digital printing systems should be aware of three limitations of these systems. They are:

- Decay of charge voltage on the photoconductor between the time of charging, exposing and toning. The amount of toner transferred to the image, which determines image contrast, depends on the exact voltage of the charge on the photoconductor at the instant of transfer.
- Toner chemistry is very complex and can cause variations in batches to toners, and high cost of formulating special toners.
- Liquid toners which are dispersed in *isopar*, a *volatile organic compound* (VOC), is subject to environmental regulations.

Manufacturers of EP systems are aware of these limitations and are developing means of avoiding them.

INK-JET DIGITAL PRINTING

Ink-jet printing technology uses jets of ink droplets driven by digital signals to print the same or variable information directly on paper without a press- or copier-like device. It has many desirable features for digital color printing.

- Ink-jet is less complex than other digital printing technologies.
- It uses much simpler devices than other technologies.
- It does not depend on light sensitivity, lasers or LEDs for imaging.
- It can produce high quality prints, has low cost of consumables and has silent operation.

Two basic types of ink-jets are (1) *continuous jet* in which drops are generated continuously and deflected to produce an image, and (2) *drop-on-demand* in which a drop is formed and emitted on response to an applied digital signal.

Continuous jets have three types: (1) charged drops are deflected electrically to form the image on the paper and uncharged drops are diverted to a gutter and are recycled; (2) uncharged drops go to the paper, and charged drops go to the gutter and are recycled; and (3) all drops are charged and are controlled by electric deflection.

AN INK-JET CONTINUOUS JET PRINTER

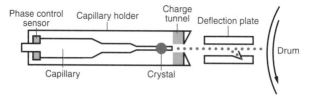

Drop-on-demand uses two types of printers: (1) *piezoelectric,* and (2) *thermal ink-jet systems:*

Piezoelectric printers use a voltage pulse applied to a piezo-electric crystal to generate a pressure pulse in the imaging head that causes emission of a droplet.

Thermal ink-jet systems have two types: (1) *bubblejet,* and (2) *solid ink/phase change.*

- *Bubblejet* printers use resistive heating that causes the liquid ink to boil. This produces an ink drop and vapor that propels the drop to the paper.

- *Solid ink/phase* change printers use heat to melt solid sticks of ink, and the phase changed melted ink is converted to drops that are ejected using impulses from a piezoelectric crystal. The drops undergo another phase change from liquid to solid on contact with the paper. The system is claimed to print on any paper with print quality equivalent to laser printers at lower cost.

Most of the color printers using ink-jet technologies have prices under $10,000, and many are well under $5,000.

Industrial ink-jet systems have been in use for printing variable information in packaging. A system has been developed for printing variable information in newspapers in single color at 1,000 feet per minute (fpm). A 4-color system has also been developed that can be printed at 200 fpm. The system uses continuous ink-jet technology with a linear array of nozzles or jets in a printhead. The technology prints multiple levels of dots at each pixel location resulting in variable dot densities that have a con-tone appearance.

Large format ink-jet printers are used extensively as proofers for checking color content, register, and imposition in CTP systems and for printing large displays for indoor and outdoor advertising.

Two serious limitations of ink-jet printing have been poor light fastness and water resistance of ink-jet inks. New inks have been developed to correct both these conditions. Also there are ink-jet inks to match *SWOP colors,* and *UV curing ink-jet* inks for printing MICR, textiles, wall coverings and commercial signs and displays. Two other limitations of ink-jet printing are slow speed and low resolution (600 dpi or lower) which are being improved.

ION OR ELECTRON CHARGE DEPOSITION PRINTING

This process uses an electron cartridge driven by digital files to produce negative charges on a drum with a dielectric surface of aluminum oxide. The image is produced by a special magnetic toner that is fixed by cold fusion. It is used only for single or spot color printing because the pressure of image transfer and cold fusion fixing can distort the substrate. The process has been used for volume and/or variable information printing of checks, invoices, forms, letters, proposals, manuals, reports, tickets and tags. Printing resolution is up to 600 dpi.

MAGNETOGRAPHY

Magnetography is a digital printing technology similar to ion or electron deposition, except that a magnetic drum is used. A magnetic charge is produced on the drum by an infinitely vari-

PRINCIPLE OF A MAGNETOGRAPHIC PRINTER

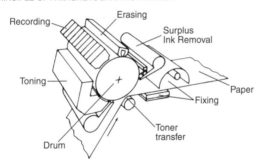

able digitally generated image, and the imaging material is a monocomponent magnetic toner. Spot and/or opaque colors can be printed but not transparent process colors, as magnetic toners are dark and opaque. Two European manufacturers make equipment for magnetographic printing using the same technology. The systems are used for printing bar codes, business forms, direct mail, labels, lottery tickets, numbering and tags. Resolution is as high as 400 dpi and press speeds up to 600 fpm are claimed.

THERMAL TRANSFER PRINTING

Thermal transfer printers use digital files of text and graphics to drive thermal printheads that melt spots of dry ink on *donor ribbons* and transfer the spots to *receiver sheets*. One commercial thermal transfer printer is used mainly for short-run label printing direct from digital files without printing plates, photoconductor or makeready. The system prints with a resolution of 300 or 400 dpi and printing speed of 20 fpm. It has a built-in computer and produces finished labels in up to four colors, laminated, butt cut, and either rotary or flatbed diecut. The labels can have over 25% variable information and can include solids, tints, logos, fixed text, sequential bar-codes, numbers and wiring diagrams.

THERMAL TRANSFER DYE SUBLIMATION PRINTING

These are thermal transfer systems in which the solid inks on the donor sheets are replaced by *sublimable dyes*. In use the thermal head converts the dye spots to gas spots which condense on the receiver. This is the configuration of the engine used in the thermal printers used for dye sublimation proofing and digital color printing.

ELECTRO-COAGULATION PRINTING

This new digital printing technology was invented and developed in Canada. It uses pigmented water-base polymetric ink sprayed on a release coating on a metallic cylinder. The image is produced on the cylinder by a digital file-controlled writing printhead that uses the cylinder as a positive electrode to coagulate the ink on same-size dots (400 dpi) in up to 256 thicknesses of ink corresponding to very short pulses of electric currents. The dot thicknesses vary according to the duration of the pulses that are controlled by the data in the digital file. After transfer of the coagulated ink to the paper and cleaning the cylinder, the imaging cycle is repeated for each revolution of the press. Two distinguishing features of the process are (1) the use of water-base ink in place of expensive toners, and (2) the use

PRINCIPLE OF ELECTRO-COAGULATION PRINTING

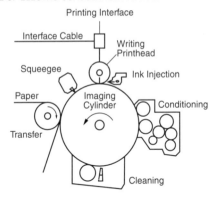

of electric impulses to produce the image instead of lasers, LEDs or other imaging media.

ECONOMICS OF PLATELESS DIGITAL COLOR PRINTING

The short-run color printing market consists of color printing in quantities, usually, under 2,000, of *variable information, on-demand* and/or *distributed printing.* Serving this market world-wide are over 5,000 digital color presses which include mainly the liquid toner single unit digital offset press, dry toner 8-unit digital web press and 4-drum digital color copier systems. For runs over 500 the digital presses are supplemented by over 1,500 direct imaging (DI) computer-to-plate on-press systems.

A major difference between plateless and plate printing systems, besides the ability of plateless systems to print variable information from impression-to-impression, is the unit cost of printed products. In plateless printing, the unit cost remains approximately the same for any quantity of prints; i.e. the cost of 100 prints is about 100 times the cost of one print. In plate printing systems, the unit cost of prints decreases as the quantities increase. The break-even point for 4-color prints of pages (one side) between plate (with on-press imaging) and plateless printing is between 300 and 500 prints. The break-even point for CTP and plateless prints is about 1,000 prints, and for conventional platemaking and printing is about 2,000 prints.

OFFSET LITHOGRAPHIC WORKFLOWS

A detailed *workflow* for conventional printing before digital imaging and printing is on pages 75-76. Briefly, it was *copy preparation, with layouts and mechanicals, line films, halftone films and color separations on camera or scanner, film process-ing, proofing, dot etching and other manual corrections, proof-ing, more corrections, contract proof, stripping, platemaking, plate processing, mounting plate on press, makeready, printing.*

Digital printing is a combination of *digital imaging* and *digital press* with an intermediate combination of *digital imaging* and *conventional press. Digital imaging* has already almost com-pletely replaced conventional prepress. This has had a very sig-nificant effect on shortening workflows.

Workflows Before the introduction of device independent Desktop Publishing (DTP) systems in the late 1980s most pre-press was done with analog scanners, electronic typesetting and device dependent color electronic prepress systems (CEPS). A typical *workflow* for prepress was: *typesetting, analog scanning, proofing, final films, film processing, contract proof, image assembly (stripping), plate exposure, plate processing, mounting plates on press, makeready, printing.*

After the introduction of the *imagesetter* in 1988 which com-bined a computer, page layout program, typographic page description language into the system called *Desktop Publishing* (DTP), the *workflow* became: *digital scanner, proofing, DTP system with color correction and page composition software, proofing, filmsetter, film processing, analog proofing, plate exposure, plate processing, mounting plates on press, make-ready, printing.*

With the latest digital imaging systems and computer-to-plate-on-press technology, the new workflow is: *digital camera or scanner, proofing, processing on DTP system with software and color management, digital proofing, plate imaging on-press, press makeready, printing.*

These are simplistic workflows. For more information on digital workflows, see page 112.

PRINTING INKS

Each printing process requires different inks. In general, all inks consist of pigments, resin vehicles in which the pigment is dispersed, solvents or other fluids to control body and other additives to induce drying and/or impart necessary working properties to the inks. Letterpress and lithographic inks are fairly stiff and require long ink roller trains on the press to get the proper flow and film thickness for printing. Gravure and flexographic inks are very fluid and dry mainly by evaporation of solvent. Screen printing inks are paint-like in consistency and drying characteristics. Toners, or digital inks, are the imaging products for digital printing and vary according to process.

INK PROPERTIES

The most important properties of printing inks are *color, color strength* and *pigment type; rheology,* including *viscosity, body, length,* and *tack; drying characteristics* and the properties of the *dried film.* Many of these properties are dependent on the process and final use.

Color – Color Strength – Pigments/Dyes

Color and other optical properties like opacity or transparency are imparted to inks by *pigments* which are finely divided solid materials and sometimes *dyes.* Some pigments such as alumina hydrate, and iron blues are inorganic but most of the pigments in use are insoluble derivatives of organic dyes. In addition to color, important pigment characteristics include specific gravity, particle size, opacity, chemical resistance, dispersibility and permanence. *Color matching* has been traditionally done by visual comparison under standard lighting conditions. Most color matching is now done with spectrophotometers and computer programs. Color matching systems using sets of basic inks, color charts and mixing formulas are offered by several companies. Also available are charts of colors that can be produced by standard process color inks like SWOP *(see page 169).*

Rheology

Viscosity is a measure of the flow characteristics of soft or fluid inks. Stiff inks can have a false body which is called *thixotropy.* Conventional letterpress and offset inks are shear thinning. They set to a fairly stiff mass in the can, but when worked on a slab with an ink knife become quite fluid and flow freely. This is a reason why so many ink rollers are needed on letterpress and lithographic presses.

Body refers to the consistency, stiffness or softness of inks. Ink consistencies vary widely from very stiff inks for collotype to very soft, fluid inks for newsprint, gravure and flexography.

Length is a rheological property associated with the ability of an ink to flow and form filaments. Inks can be *long* or *short*. Long inks flow well and form long filaments. They are undesirable, especially on high-speed presses because they have a tendency to *fly* or *mist*. Newsprint inks are characterized by this property. Short inks have the consistency of butter with poor flow properties. They have a tendency to pile on the rollers, plate or blanket. Most good inks are neither excessively long nor short.

LONG INK   SHORT INK

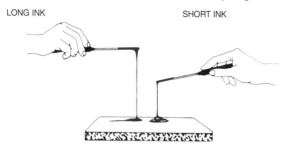

Tack is the *stickiness* of the ink, or the force required to split an ink film between two surfaces. It is an important requirement in the transfer of ink in the ink train to the plate and then to the paper in letterpress or to the blanket and paper in offset. Tack also determines whether the ink will pick the surface of the paper, will trap properly in wet multicolor printing, or will print sharp, clean lines and halftones. If the tack of the ink is higher than the surface strength of the paper, the paper will pick, split or tear. In wet multicolor printing, as in letterpress and lithography, the first ink down must be tackier than the next ink at the instant of transfer, or the second ink will not transfer to (trap on) the first color. Offset inks must be tacky to print sharp images and resist excessive emulsification with the fountain solution. Compromises must be reached when jobs contain both solids and halftones since tacky inks do not print smooth solids. Ink tack can be measured on an Inkometer or Tackoscope.

Drying

Drying of inks is important because a printed piece cannot be handled or used until the liquid or plastic ink film has solidified and dried. Printing inks dry in a number of ways: absorption, selective absorption, oxidation, polymerization, evaporation, precipitation and curing by radiation. Inks can be formulated to dry by a combination of two or more of these mechanisms. The first stage in drying is setting, and often this is more important in printing than the actual drying.

New systems for drying or curing inks have been developed to eliminate pollution caused by the evolution of solvents and other effluents associated with ink drying. Inks that cure by UV and/or electron beam (EB) radiation are described on page 161. Infrared units have been developed for drying special inks of the low-solvent super quick-setting type. Inline overcoatings are used to eliminate anti-set-off sprays and assist drying in sheetfed printing *(see page 164)*.

Job Planning

When a job is first planned, "thinking ink" can avoid a lot of head-aches and expense later. Remember that the inkmaker must make the ink for the substrate which is mostly paper. Ink and substrate relationship is very critical. Samples of substrate to be printed should be furnished when ordering inks. If this cannot be done, give the description and specification for the substrate. In addition, the inkmaker should know the printing process, color rotation, type of press, press speed, drying demands, gloss, varnish and any special requirements such as alcohol-resistance, fade-resistance, etc., that are necessary for the final product performance *(see Special Characteristics, page 164)*.

LETTERPRESS INKS

Letterpress inks are designed for printing from raised surfaces such as type, engravings and electrotypes. These inks are usu-ally of moderate tack and viscosity. Most sheetfed letterpress inks consist mainly of pigments ground in a drying-oil vehicle with driers added. They dry by oxidation. They may also contain special resins and compounds to give characteristics such as gloss, scuff-resistance, etc. Letterpress inks for web printing dry by penetration *(newsprint inks)*, by evaporation *(heat-set inks)* or by precipitation *(moisture-set inks)*. Heat-set inks are formu-lated with high-boiling, slow-evaporating petroleum oils and sol-vents. These provide maximum press stability, yet dry rapidly with heated air applied in a drier.

OFFSET-LITHOGRAPHIC INKS

Lithographic inks are formulated to print from planographic sur-faces, which use the principle that grease and water do not mix. Lithographic inks are generally very strong in color value to com-pensate for the lesser amount applied. They are the strongest of all inks next to collotype inks. The average amount of ink transferred to the paper is about half that of letterpress.

Basically, sheetfed lithographic inks are similar to letterpress inks of the oxidizing type. However, they contain more water-resistant vehicles and pigments that do not bleed in water or alcohol that is used in fountain solutions. Heat-set litho inks are also similar to their letterpress counterparts but use special vehicles and pigments to resist reaction with water and chemicals that are used in lithographic fountain solutions.

The development of new vegetable oil vehicles has improved performance especially of news inks. It is a renewable resource which helps make the industry independent of petrochemical oil vehicles. Linseed and rapeseed oils are vegetable oils that have been used for years. The largest volume vegetable oil is *soybean oil* used to manufacture *soy news ink*, which has good printing qualities and reduces smudging from newsprint.

Letterset inks are used on offset presses with raised image plates and without dampening solutions. They have no pigment restrictions as no dampening solution is used. They are stronger than letterpress, but not as strong as conventional offset inks because the inks are not diluted by emulsified water during running on the press.

Waterless inks are like letterset inks, but are specially formulated to resist heat effects that can cause toning of non-image areas of waterless printing plates. These plates have silicone compounds on the non-image areas which resist wetting by ink without the need for dampening solution. They are, however, very temperature sensitive, as a rise in ink temperature of over 10° F can cause toning in the non-printing areas. Most presses used for waterless printing have temperature controlled inking rollers. A waterless ink / waterless plate system has been developed to facilitate press operation.

Single fluid inks have been developed to run with standard lithographic plates on lithographic presses without the use of dampening systems. They have been tried for years without success. The newly developed inks show promise as a solution to the difficulties associated with the use of fountain solutions. If successful, they could compete with waterless inks.

GRAVURE INKS

Gravure inks are rapid-drying fluid inks that have the proper viscosity to be drawn by capillary action from the engraved wells in the cylinder or plate. They dry principally by the evaporation of the solvent in the ink, with or without the use of heat. Gravure inks must be free of abrasive particles that could wear the engraved cylinder or plate.

A wide variety of solvents are used in gravure depending on the resin system used. Most gravure solvents are very volatile and can cause fires or explosions if not handled properly. Solvent recovery systems are used in many publication plants to eliminate pollution from the evaporated solvents. Water-based gravure inks have been developed mainly for packaging to eliminate both the fire hazard and solvent pollution.

FLEXOGRAPHIC INKS

Flexographic inks are fast-drying fluid inks similar in viscosity to gravure inks. They are used in printing a large variety of surfaces, from carpeting and wallboard to cellophane and plastic films, metal foils, etc. The exceptional color effects afforded by flexography are best exploited by using large masses of color. Flexo inks consist of colorants, which may be either pigments or soluble dyes, together with a vehicle or binder and volatile solvents. They are usually alcohol- or water-based although other solvents are also used. Alcohol-based inks are the most common and dry by evaporation. Water-based inks cost less and dry by both evaporation and absorption on paper. Water-based flexographic inks are used in flexo newspaper printing because they are practically smudge proof when dry. UV flexographic inks are responsible for many improvements in image quality.

A new development in flexography is the use of very fine anilox rolls (500 to 1,100 lines/inch) which require the use of very strong water-base flexographic ink. Some of this printing is difficult to distinguish from lithography. Also, some of the pigmentation is as strong as in lithography.

SCREEN PRINTING INKS

Screen inks are usually of the drying oil or resin solvent types, and have the consistency of thick paint. Inks are made in any color, using a suitable binder for the material to be printed. All inks are short and buttery to print sharp and squeegee with little resistance. To prevent clogging of the screen, the solvents used should not evaporate too rapidly. Clear UV screen inks are used extensively for coatings.

TYPES OF INKS

Many types of inks are made for different printing uses. Some inks have been developed especially to eliminate starch anti-set-off spray in sheetfed printing and air pollution from heat-set ink solvents in web printing. There are inks that simulate metallic luster;

inks that print magnetic characters which can be read on special electronic equipment; inks that are alcohol and scuff-resistant for liquor labels; inks that are alkali-resistant for soap packages; and inks that have high brilliance for attractive displays.

Radiation curing inks have been developed to dry instantly. This can eliminate spray powder in sheetfed printing and air pollution from solvents in conventional web heat-set inks. There are two types of these inks: ultraviolet (UV) and electron beam (EB) curing. Radiation curing inks must be handled properly due to their potential dermal toxicological effects.

UV curing inks consist of pigments dispersed in liquid prepolymers (oligomers), monomers and initiators which on exposure to intensive UV radiation release free radicals that instantly polymerize the vehicle to a dry, tough thermosetting resin. Because the active ingredients in these inks are more costly than the solvents and vehicles they replace in conventional inks, they are more expensive (as much as three times the cost of regular inks). Therefore they are used mainly where improved film performance or instant cure is needed. They have found use in luxury packaging such as liquor and cosmetic cartons, for metal decorating, screen printing and coating, and more recently for letterpress and flexographic printing. UV inks on narrow web presses have revived the use of letterpress in some printing markets such as pressure sensitive labels. The use of UV flexo inks have resulted in significant improvements in image quality such as the printing of 3% highlight dots on 150 lpi screen images.

Electron beam (EB) curing inks make a good alternative to UV inks since no expensive initiators are needed and some lower cost, less reactive materials can be used. The major disadvantage of EB is the high capital cost of equipping a press to use it. A reason is the required use of nitrogen as an inert blanket because oxygen must be reduced to less than 200 ppm. Despite these limitations there are a number of successful commercial installations of EB on presses. EB uses less energy than UV, which in turn uses about half the energy of gas drying.

Quick-setting sheetfed inks have been very successful for printing on enamel and cast-coated papers in both letterpress and lithography. These inks consist of a delicately balanced oil-resin-solvent vehicle system. Upon contact with the paper during printing, the paper surface or coating quickly drains some of the solvents from the ink, yielding a film that sets rapidly and permits handling. The ink film then continues to dry by oxidation. Quick-set inks usually dry with a good gloss.

Super quick-set infrared inks are a modification of quick-set inks using new synthetic resins with controlled solubility properties used in combination with special blended solvent systems and a minimal amount of drying oils. The setting of these inks is accelerated by the application of heat energy, and IR radiation is a convenient way to apply this energy on a sheet-fed press. Even without heat, these inks set considerably faster than the conventional quick-set inks which they replace.

Heat-set inks are quick-drying inks for web publication printing. The solvents used are evaporated from the ink film as they pass through a heating chamber (dryer), leaving the pigment and binding resins fixed to the paper in such a manner that there is no chance for ink spread or excessive penetration into the paper. Presses must be equipped with a heating unit, a recirculation afterburner or absorption system to prevent air pollution and chill rolls to cool the heated resins and set the image on the web so the web can be cut and folded.

High-gloss inks basically contain special resins and varnishes that produce a glossy appearance when dry. For best results, stocks specially coated for gloss inks should be used. In general, the more resistant the paper is to penetration of the vehicle, the higher the gloss. This property of paper is generally called *holdout*. Heat used in drying can reduce the gloss of some inks. High-gloss inks are available for most printing processes.

Metallic inks use metallic powders, such as aluminum and bronze powders, mixed with the proper varnish base, to print images with pleasing metallic luster. This is because the powders are actually flakes which deposit in reflective layers. The bronze powder and vehicle for preparing gold inks are mixed just before using, since the majority of gold inks tarnish rapidly after mixing. The varnish used dries rapidly and has sufficient binding qualities to hold the powder to the paper surface. Coated papers give the best results. On rough surface papers, a base ink is usually printed first, allowed to dry and overprinted with gold. Both aluminum and gold inks can be printed by letterpress, offset or gravure. When printed by offset, alkaline or neutral fountain solutions are used to prevent tarnishing of the bronze powder by acid solutions.

Moisture-set inks are letterpress inks consisting of pigments dispersed in a vehicle composed of a water-insoluble binder dissolved in a water-miscible, or water-receptive solvent, usually a glycol. Upon subjecting the printing to either steam, fine mist or water, the water-miscible solvent absorbs some of the water which causes the resin to precipitate and bind the pigment firmly to the paper. Moisture-set inks are relatively free from odor, making them ideal for food package printing. Moisture-set inks have largely been displaced by water-based flexographic inks.

Magnetic inks were developed to increase the speed and efficiency of handling bank checks. These inks are made with iron oxide pigments that can be magnetized after printing, and "recognized" by electronic reading equipment. These inks must be formulated to produce high-grade print characteristics that will meet the rigid requirements of the reading equipment. Makeready and amount of ink must be precise and consistent.

1 2 3 4 5 6 7 8 9 0 ⑊ ⁚⁚ ⁚⁚ ⁚⁚

Each of the 10 numbers and four symbols shown above has a distinctive shape which can feed information to a computer that processes the information for a number of uses.

Fluorescent inks were formerly limited to screen printing. New smaller particle size pigments and greater pigment strength now permit colors to be printed with significant color strength in one impression by letterpress, lithography and gravure. Duotones and even full-color process are now feasible. The naturally bright inks reflect and emit light, making use of ultraviolet light waves which other inks cannot utilize. The semi-transparency of the inks permits overprinting to achieve color mixture. The fluorescent pigments are not light-fast.

Fluorescent inks must be printed on a white surface. They provide maximum brilliance when contrasted with dark surrounding hues. They are used for jobs of a semi-permanent nature, such as labeling, packaging and direct mail. Fluorescent pink is used as a fifth color in 4-color process printing to enhance skin tones, and extend the range of magenta hues in the images. Fluorescent yellow and magenta images are used as additional colors in some of the new 6-color *Hi-Fi* printing systems.

Oil ink/water washup system has been developed which uses vegetable oils that print like oil base inks and acid fountain solutions (pH 4.0 to 5.5) but are water miscible at pHs above 9.0. When printing is completed, the press rollers and blankets are washed with a water base solution (pH 9.0-10.0). The system is free of volatile organic compounds (VOCs). These inks have found limited application because many pigments cannot be used with the high acid-value vehicle. They are used extensively in check printing.

Water-based overcoatings are used in some sheetfed printing to replace off-line varnishing and eliminate the need for anti-set-off starch sprays, which are a scourge in the pressroom. Acrylic type emulsions with water and alcohol and varying degrees of gloss are coated over the wet inks on the image inline with the printing. The resin coats the ink, while the water or alcohol disperse in the paper. The coatings dry rapidly preventing the wet inks from scuffing or marking while they dry normally. UV-curable clear coatings are also used as overcoatings on the printing inks. The inks may need reformulation to be compatible with the UV overcoatings. They work best over UV inks.

A disadvantage of press overcoating has been the need for an additional unit on the press to apply the coating. The use of overcoatings has become so popular, however, that most new sheetfed presses can be supplied with special coating units or *towers* for controlled inline application of the coatings.

Varnish and lacquer are used as coatings over printing to protect the printing and increase gloss. Inkmakers should know when printing will be overcoated so that inks can be formulated properly. Otherwise, the inks are apt to bleed through the varnish or lacquer. Inks to be varnished or lacquered should not contain waxes which can prevent wetting, or adhesion of the varnish or lacquer to the ink. Also, minimal spray powder should be used on sheets to be varnished or lacquered since the powder can affect even transfer of the varnish or lacquer. The varnish maker should be notified of the chemical resistances (soaps, acids, alcohol, etc.), scuff tests, gloss requirements and other general specifications required of the varnish so the proper formulation can be produced.

Lacquers are applied on special coating machines. Most varnishes are applied on-press, from a blank or imaged plate inline with the printing on the press, and drying is by oxidation without heat. Gloss and other special characteristics are limited as the varnishes must be compatible with the wet inks.

SPECIAL CHARACTERISTICS

Inks must have other special characteristics to be satisfactory for the variety of uses to which printed matter is subjected.

- Inks must dry so that they are *rub* and *smudge-resistant*.

- Labels and packaging printing must be *scuff* and *scratch-resistant*.

- Printed matter for window displays and outdoor use require inks that are *light-fast* and *fade-resistant*.

- Ink used for soap wrappers must be *alkali-resistant* and not bleed with the product.
- *Alcohol-proofness*, or resistance to smearing by alcohol, is a must for liquor labels.
- Wrappers to be hot waxed must have inks that do not bleed in paraffin.
- Inks must be specially formulated for overcoating, especially by UV coatings.

TONERS / DIGITAL INKS

Just as each conventional plate printing process requires different inks with special requirements, each digital plateless printing process uses different materials called toners with special characteristics. The first toner was developed for electrophotography in 1938. Until 1978 when the first laser printer was introduced, toners were used entirely for copying. Since the first 135 ppm laser printer was introduced in 1989 and two digital color presses in 1993, the market for toners has exploded. In 1996 more copies were made on laser printers than on copiers. The market has expanded so much that manufacturers of digital presses, printers and toners prefer to call toners *digital inks*.

The more important toners/digital inks are for electrophotographic (EP) and ink-jet digital printing systems as these are more widely used than ion or electronic deposition, magnetography, thermal transfer, dye sublimation or electro-coagulation systems.

EP systems can use either dry or liquid toner/ink systems. Liquid systems have higher resolution than dry systems as the pigment particle size in liquid systems is much smaller. The toner, however, is dispersed in isopar, which is a volatile organic compound (VOC) and subject to federal regulation. Dry toners/inks are either dual component using special magnetic carrier beads, or monocomponent. Over 90% of EP copiers and laser printers use dual component toners.

Ink-Jet systems use several types of inks. Waterbase inks are used in continuous jet, piezoelectric drop-on-demand and bubble-jet systems. Solid inks are used in thermal phase change systems. Most ink-jet inks use soluble dyes which are not light-fast and fade on exposure to UV in sun rays. Heavy coverage of water base inks affects paper surfaces so special coating treatments are required for quality printing. New light- and water-fast inks have been developed as well as UV curing inks.

New jet inks using pigments in place of dyes have been developed for printing on textiles and paper. These inks are made with very finely ground pigments with particle size small enough to pass through the fine nozzles of most ink-jet printers. These inks have permanence associated with conventional inks.

Ion or **electron charge deposition** systems use special magnetic toners.

Magnetography uses special monocomponent magnetic toners.

Thermal transfer systems use special thermoplastic resin-based inks coated on donor ribbons.

Dye sublimation systems use donor ribbons coated with inks containing special sublimable dyes.

Electro-coagulation systems use special water-base inks with resins that can be coagulated or precipitated by the iron ions released by the short electric pulses used in the process.

ENVIRONMENTAL AND ENERGY CONSIDERATIONS

Governmental regulations severely restrict all volatile organic compounds (VOCs-solvents) commonly used in printing inks from being emitted to the atmosphere without some control method such as incineration or solvent recovery. The regulations on permissible emissions are so stringent that printers who cannot recapture solvents are required to purchase incineration equipment. Especially affected are gravure and flexography. There are also regulations on the discharge of washup solvents in sewers and disposal of solid waste.

Enforcement of such regulations could stimulate development and use of new radiation-cured, chemically-reactive, or water-based inks, but timing of the regulations and printers' investment in expensive anti-pollution devices have delayed such developments.

Increasing costs and decreasing availability of energy sources from fossil fuels such as natural gas and oil will direct attention of ink manufacturers to renewable sources of ink vehicles such as vegetable oils, for example, soybean, linseed and rapeseed oils, and alternative methods of ink drying which are more energy efficient. The dominant printing processes of the future will be those that use imaging materials (ink or toner) that are renewable, non-toxic, non-flammable and non-polluting, and at the same time reasonable in cost.

STATISTICAL PROCESS CONTROL

Printing is a manufacturing process. Since machinery wears, and variable products such as paper, inks, toners, press blankets, fountain solutions, plates, films, etc. are used, every effort must be made to control them. The system by which the printed product is controlled is called Statistical Process Control (SPC, formerly called *quality control).* Successful SPC depends on four functions of *quality assurance:* (1) specifications and control of raw materials, (2) control of the printing process, (3) standards and tolerances of acceptability and (4) inspection.

RAW MATERIALS

The printing processes use many raw materials that should be controlled. Printers should establish realistic specifications for the raw materials they use. Important material properties are *paper characteristics* such as moisture content, gloss, brightness, ink absorption, piling and picking tendencies; *ink characteristics* such as tack, yield value, gloss, drying time, fineness of grind; *fountain solution characteristics* such as pH and conductivity; *roller characteristics* such as composition and Shore hardness; *lot numbers* of photographic films and presensitized plates, and digital plates and inks.

PROCESS CONTROL

To control the printing process, instruments, targets, color bars, grayscales and other devices are needed to provide objective measurements and numbers to characterize the process. The most important instruments for checking color printing processes are *densitometers* and *spectrophotometers.*

Densitometers measure optical density, or relative degree of light absorption or opacity of an image. The darker the image, the more light it absorbs and the higher its optical density. The densitometer does not measure color as the eye sees it but as the photographic emulsion responds to it. *Transmission densitometers* are used to measure color transparencies, film negatives and positives and control the photographic process.

Reflection densitometers use red, green and blue color filters to measure and control color proofs and sheets printed with special color bars, consisting of small blocks of each color, graded halftone tints and overprints of the colors. Measured are ink strength (color solids), dot gain (color tints), neutral gray balance and ink trapping (color overprint) on the press. The readings are useful as long as the same pigments, paper and press are used. When materials are used that change these readings the densitometer is not a reliable control instrument, and a *colorimeter* or *spectrophotometer* should be used.

Colorimeters record colors as the eye sees them. They use different filters than the densitometer, and the readings can be used to describe the color in any color space or system, such as RGB (red, green, blue) which are the colors of color separations and the video screens of color workstations, and CMYK (cyan, magenta, yellow, black) which are the printing colors. New color sensors have made possible new colorimeters with high speed and precision in portable instruments about the size and shape of densitometers at reasonable cost.

Spectrophotometers measure colors in narrow bands (usually 10 nm wide) across the visible spectrum and draw a curve of the color. There are new combination portable spectro-colorimeters whose readings can also be converted to densitometer readings.

Control targets besides gray scales and color bars are used as controls to standardize the process. *Sensitivity guides* are continuous-tone gray scales with numbered steps which are used to control exposures in platemaking and lithfilm photography. *Star targets* are pinwheels about ½" in diameter, which are used to measure image resolution during plate production, and plate degradation, dot doubling, dot gain and slurring during printing. The UGRA Test Target is a measure of image resolution and dot size on plates and in printing. More than 50 test targets are in use worldwide. All are designed for image control in printing and are equally useful depending on the experience, interpretation and judgment of the operator.

GATF SENSITIVITY GUIDE   GATF STAR TARGET

PARTIAL UGRA TEST TARGET

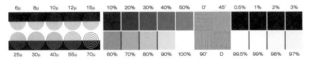

Quality control instruments besides densitometers and colorimeters are used to check raw materials and the stability of the printing process. *Paper hygroscopes* are hand-held, sword-shaped instruments with hygroscopic elements used to check the equilibrium moisture balance of paper with the relative humidity of the pressroom to ensure printing without distortion, wrinkles and misregister. *Inkometers* are instruments with temperature controlled rollers and calibrated ink film thicknesses used to measure ink tack and length, which affect the trapping characteristics of inks.

Fineness of grind gauges are steel blocks with two precision ground calibrated wedge grooves with graded depths from 0 to 0.001″ used with a scraper blade to determine the presence of improperly ground ink pigment particles. *Ink film thickness (IFT) gauges* are wheels with hubs and precision ground grooves like the fineness of grind gauge to measure the IFT on the steel roller above the ink form rollers. IFT is extremely important for consistent printing as it affects dot gain and plate wear. *Blanket packing gauges* are devices with micrometers used to control press packing by measuring plate and blanket height above or below the cylinder bearers. *pH and conductivity meters* are electronic instruments used to measure and monitor the press fountain solutions during printing. Special precision 10- to 20-power *magnifying glasses* are important for the examination and analysis of printing defects and their causes.

STANDARDS AND TOLERANCES

Process standards and tolerances vary with the type of product and the quality levels demanded. Some customer groups such as the Magazine Publishers Association (MPA) have established standard ink colors and dot gain tolerances for web offset publication printing known as SWOP. Register and color variation are two of the most critical characteristics requiring definition and specification. Commercial and hairline register are discussed on page 58. Acceptable *color variation* is generally ±0.02 density units for yellow, magenta and cyan, and ±0.04 units for black.

INSPECTION

Periodic inspection is necessary in any SPC system to make sure the printing system is functioning properly. *Quality cannot be inspected into printing jobs.* Unless there is strict adherence to raw material specifications, rigid control of process variables and realistic standards and tolerances, no amount of inspection will improve quality and reduce waste. When these facets of quality control are properly set and used, satisfactory printing becomes a matter of course, and periodic inspection is needed mainly to be sure the system is working.

COLOR MANAGEMENT SYSTEMS

A major difference between *device dependent color electronic prepress systems (CEPS)* and *device independent desktop color publishing (DTP) systems,* is the different *color spaces* used. CEPS performed all color image processing in a CMYK color space so calibration was limited to the measuring instruments. DTP, on the other hand, does all the digital imaging (image capture on scanner or digital camera) in an RGB color space which must be converted to a CMYK color space for printing. This conversion involves intensive color management and calibration to assure that the RGB values of the digital imaging systems always convert to the required CMYK values in the printing systems.

International Color Consortium (ICC) is a primary solution established in 1993 by a number of suppliers for the purpose of creating, promoting and encouraging the standardization and evolution of open, vendor-neutral, cross-platform color management system architecture and components. The new standards formulated by the ICC group for color management (CM) were designed to allow color consistency between different devices, applications and platforms. The ICC concept depends on the use of device profiles that are workable on any platform.

A main reason CM systems are working is the world-wide acceptance of *color targets* for scanning (RGB color spaces) and printing (CMYK color spaces) embodied in the ANSI standards IT8.7/1 (transmission) and IT8.7/2 (reflection) scanner (RGB) targets, and IT8.7/3 printing (CMYK) targets. These targets are used to produce scanner and printer profiles and are available from NPES (Reston, VA 22091). ICC profiles supplied by many scanner and printer manufacturers are for average products and not a specific device. For this type of profile, users need to produce their own using special software products produced by digital imaging suppliers.

Postpress

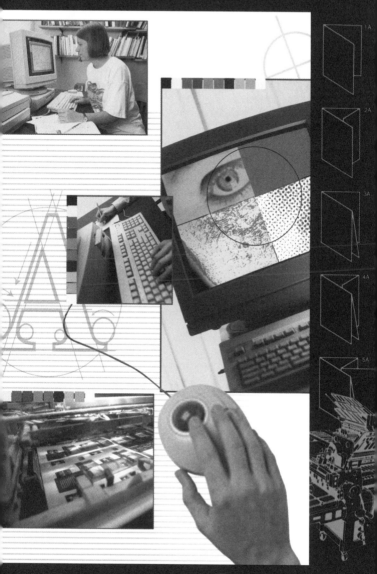

A printed job is not complete until it is converted to a finished marketable product and delivered. These are the functions of *postpress,* which consists of the supplementary operations to printing: binding, finishing and distribution. Some printing like stationery, small posters, notices, etc. can be delivered as printed, but most printing must be converted from printed sheets to a finished printed piece, through various binding and finishing operations.

The work required to convert printed sheets or webs into books, magazines, catalogs and folders is called *binding.* The operations to make displays, folding cartons and boxes, tags, labels, greeting cards and a variety of special packaging and advertising materials are known as *finishing.* The processes, facilities and means used to handle, package and ship printed products to the customer are known as *distribution.*

New trends in finishing and binding are automated systems for material handling with new conveyor systems using Print Rolls, new inline binding equipment with automated makeready, and new binding equipment for digital presses.

BINDING

PAMPHLET BINDING

This is a general term for binding folders, booklets, catalogs, magazines, etc., as opposed to bookbinding which will be discussed later. There are generally five steps in pamphlet binding: *scoring, folding, gathering* or *collating, stitching, cutting* and *trimming.* Most printing requires one or more of these, but not all. For example, a printed folder is trimmed to size and folded. When printed sheets are delivered to the bindery, the first step is to fold the sheets (most commonly in multiples of fours) into sections or signatures. In the case of heavyweight or cover paper, folding is made easier by first scoring.

Scoring

A score is a crease in a sheet of heavyweight or cover paper to facilitate folding. As a rule, only those methods which produce

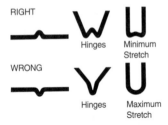

RIGHT — Hinges — Minimum Stretch

WRONG — Hinges — Maximum Stretch

an embossed ridge on the paper will give good folding results. The fold should always be made with the ridge or hinge on the inside for minimum stretch *(see illustrations)*. Booklet or catalog covers must have a score wide enough to take the necessary number of pages without strain on the fold.

The most common method of scoring is using a round face scoring rule locked in a form on a platen or cylinder press. The width of the rule varies with the thickness of the paper. A thicker paper requires a thicker rule which will give a wider crease to help make a cleaner fold.

Folding

Paper is usually folded on a *buckle* or a combination *buckle/ knife* machine. Contemporary folding machines are set up by computer in seconds. These systems offer several hundreds of different folding combinations. The sheet is carried on conveyor belts from an automatic feeder, and rollers force the sheet into a fold-plate, which is adjustable to the length of the fold. The sheet hits a stop in the fold-plate, buckles, and is carried between two other rollers which fold the sheet. There can be as many as 64 pages to a signature.

FOLDER

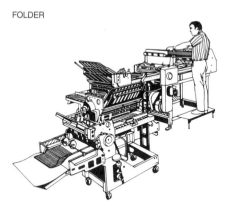

There are two kinds of folds: *parallel* and *right angle*. In parallel folding, each fold is parallel to the other. An example is a letter which requires two parallel folds for mailing. An *accordion* or *fan fold* is a type of parallel folding used extensively for computer printout forms. A right-angle fold is two or more folds, with each fold at right angles to the preceding one. Folding machines can be equipped with attachments for scoring, trimming, slitting, perforating and pasting.

In designing printing, the different types of folds and the limitations of mechanical folding should be considered at the planning level. Otherwise, one or more folds might end up being a costly hand-folding operation. The sketches below illustrate the most common types of folds.

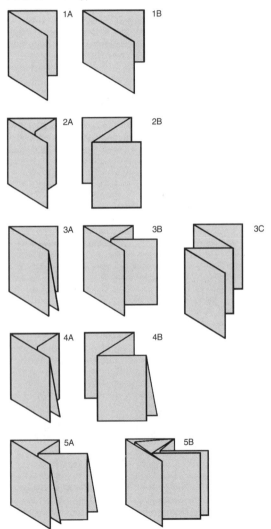

Types of Folders

1. Four-page folder Simplest type of folder, with only one fold, folding either on the (A) long or (B) short dimension. Used for bill stuffers, instruction sheets, price lists, etc.

2. Six-page folder Made with two parallel folds, either (A) regular or (B) accordion. Used for letters, circulars, envelope stuffers, promotional folders, etc.

3. Eight-page folder Illustrated in three ways, (A) one parallel and one right-angle fold, also called *french fold* when printing is on one side of the paper, (B) two parallel folds and (C) three parallel accordion folds, for ease in opening. Also, (A) and (B) can be bound into an 8-page booklet, (C) cannot.

4. Twelve-page folder Illustrated in two ways, both with one parallel fold and two right-angle folds, either (A) regular or (B) accordion.

5. Sixteen-page folder Shown in two ways, (A) one parallel and two right-angle folds and (B) three parallel folds, used for easy-to-open transportation schedules. Also, can be bound into a 16-page booklet.

Collating/Gathering

Single sheets are collated; folded sheets are gathered. Most binders use sophisticated electronics which read every signature to give 100% assurance that all signatures are gathered and in right sequence. High-speed lines use print rolls — signatures that are rolled up at the press and unwind at the binding machine. Collating can be done by hand or machine, depending on the size of the job.

Saddle stitching

After the signatures are collated, they can be stitched together.

SADDLE STITCHER

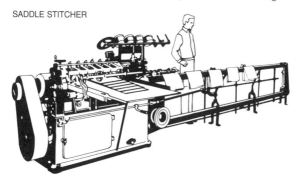

The booklet is placed on a saddle beneath a mechanical stitching head, and staples are forced through the backbone or spine of the booklet. This type of binding is the simplest and most inexpensive. Booklets will lie flat and stay open for ease in reading. Most booklets, programs and catalogs are saddle-stitched.

SADDLE-STITCH

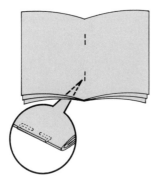

Cutting

Computer controlled and programmed guillotine cutters have a single knife and are used to cut labels, paper, etc. as required. Sophisticated systems feature automatic loading and unloading resulting in high production.

PAPER CUTTER

Trimming

Trimming is usually done with three knives in a single cycle. Modern computer-programmed trimmers can be set-up or changed over in minutes. Trimmers are most often inline on high-speed equipment or off-line in a digital, on-demand printing and binding environment.

BOOKBINDING

There are many ways to bind a book, but the most common methods are *edition binding*, also known as hardcover or case binding (best when permanence is required), *perfect binding* (widely used for inexpensive paperback books) and *mechanical binding* (for manuals and notebooks).

Edition binding There are basically two methods used to bind the bookblocks together: sewing and adhesive binding. In sewing the individual signatures are sewn through the fold into a bookblock and thereafter receive an adhesive coating on the backbone. These books open flat and offer the best durability.

CONSTRUCTION OF A SEWN OR ADHESIVE BOUND BOOK

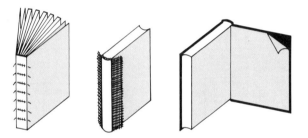

Adhesive binding employs many techniques using one- and two-shot adhesives such as Burst, Notch or Threadseal bindings. Difficult-to-bind paper stocks are bound with PUR (Poly Urethane) adhesives. All receive a four-page endpaper or may feature self-contained endpapers, which are part of the paper stock. Paper grain must always be parallel to the binding edge or covers will warp.

Thereafter, the sewn or adhesive bound books are trimmed. Some books are rounded and backed, others are flatback. Modern edition binding lines can be changed over in minutes to accommodate new publishing trends to print and bind on demand.

Perfect binding is used for softcover bindings. The sheets are held together by a flexible adhesive. After gathering, the signatures are prepared by grinding off approximately ³⁄₁₆″. This process will expose the paper fibers which then are imbedded into an adhesive. The covers are attached inline. The *Pocket Pal* is an example of perfect binding with hotmelt adhesive.

High quality adhesive bindings use one- or two-shot systems: PUR (Poly Urethane) or cold emulsion adhesives. LayFlat bindings such as Otabind and the cloth reinforced Repkover are very much in demand as these bindings do lie flat and are much more economical to produce than mechanical bindings.

Mechanical binding is used for notebooks and other types of books which must open flat. The sheets are punched with a series of round or slotted holes on the binding edge. Then wire, plastic coils or rings are inserted through the holes. Looseleaf notebooks are a form of mechanical binding with rings which open to allow removal or addition of pages. In designing a book for mechanical binding, allowance must be made in the gutter (inner margin) of the book for the punched holes.

PLASTIC   SPIRAL

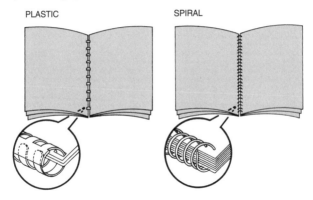

FINISHING

Finishing is a general term that includes a number of different operations and specialties. For instance, mounting, die-cutting and easeling of displays; varnishing, laminating, embossing, bronzing, die-stamping, pebbling, marbleizing, dyeing, stripping, folding, collating, perforating, punching, round cornering, padding and tin-edging of printed and unprinted materials; cutting, creasing, stripping and gluing of folding paper cartons; or slotting and gluing corrugated boxes. Most of these operations are highly specialized.

Some finishing operations are performed inline with printing on web-fed presses. In newspaper, magazine and book printing, folded signatures are delivered from the press. In some types of packaging, particularly on flexographic presses, the cutting, creasing and stripping are done inline with the printing. The new UV inks make possible inline operations like lacquering, folding, slitting, etc., in sheetfed printing.

Two finishing operations quite often used are embossing and die-cutting.

Embossing The image is molded in embossing so that it is raised in relief. Molding is achieved by pressing the material to be embossed between a brass female die and a male bed or counter mounted in register on a press. Strawboard, plastic, molding compound or newspaper matte material may be used for a counter. Light embossing may be done without heat on a cylinder or platen press. For heavy embossing and where fine detail is required, the die is fastened to a heated plate on a heavy-duty arch or *four-post* press. Embossing may be done in register with printing or on blank stock giving a bas-relief effect. The latter is called *blind* embossing, and a soft paper is best. If a metallic effect is desired, special stamping foil is used.

Diecutting There are two methods in use today: high or hollow diecutting and steel-rule diecutting. Hollow diecutting is a process used almost exclusively for labels and envelopes. A steel die, which is hollow like a cookie cutter, is positioned on a jogged pile of printed sheets. Pressure forces the die through the pile. The labels remain in the die until stripped out by hand.

Steel-rule diecutting is used for larger size dies or where close register is required. The dies are hardened steel rules bent to a desired shape and inserted into jigsawed gaps in ¾″ plywood *dieboards*. The multiple dies are locked up in a chase. They are positioned and made ready on a platen diecutting press. Flatbed cylinder presses are also used for diecutting. The ink rollers are removed to avoid being cut, and a steel jacket is secured around the cylinder. Dies must have small nicks to prevent the diecut area from dropping out while on press.

The latest developments in diecutting are laser-cut dieboards and rotary diecutters. In the laser cutting of dieboards, CAD/CAM techniques are used to lay out the designs on the boards which are cut by high-powered lasers. Rotary diecutters are expensive and are used mainly on very long runs at high speed as an additional inline operation on web presses for specialty products like milk cartons.

Shrink-packaging is used to wrap printed pieces. In addition to being able to see the product through the wrapping, its advantages are increased production and reduced labor cost.

The equipment is simple to operate. The product is inserted into a folded roll of polyethylene film which is heat-sealed around the product. It then goes into the shrink tunnel where the proper temperature shrinks the film tightly around the product.

BINDERY AUTOMATION

Technological developments in bindery equipment have made available: automatic units for counting, bundling and material handling; automatic hopper loaders and bundle distribution systems; automatic palletizers and depalletizers; microprocessor-controlled cutting machines with automatic jogging, loading and unloading; folders with increasing speed and accuracy; automated adhesive binding systems; increased inline processing and computerized on-press addressing using ink-jet printers. In addition, robotics and artificial intelligence are advancing bindery automation with the use of automated guided vehicles (AGV) for delivering paper rolls to the press and moving printed signatures to the bindery.

ON-DEMAND BINDING

Digital printing has made many advances but it will not advance much further without appropriate on-demand binding systems. Softcover binding is now often done inline. Better bindings are achieved off-line using cold emulsion adhesives and special Repkover cover structures which offer lay-flat features.

There are also many mechanical bindings using conventional systems. In addition, hardcover bindings are now possible using specially prepared book covercloth which can be printed on digital presses. With such a variety of contemporary features, soon anyone can become a publisher, printing and binding just a few or a few hundred books.

DISTRIBUTION

Mailing and distribution are important functions recently added to postpress operations. Most present methods of distribution are inadequate. In many cases the cost of distributing a printed product can exceed 25 percent of the total cost of publishing and printing it! Considered in mailing and distribution are: (1) providing customers with mailing expertise and establishing a post office on customer premises when necessary, (2) use of computerized ink-jet systems for addressing, and (3) development of advanced mail sorting and handling systems, such as combined mailing, co-mailing/pool mailing and commingled mailing.

Paper

Since paper, paperboard or other stock on which an image is printed can represent 35–55 percent of the final cost of the printed job, a working knowledge of paper assures that the stock selected is the most cost-effective grade needed to meet the demands of a particular print job. The paper chosen should have the desired printability and runnability to ensure optimum results with minimal problems. Everyone involved should know as much as possible about the manufacture and characteristics of paper, since these can have a significant bearing on the appearance of the job, and the printer's ability to print it.

PULPING

The first step in papermaking is the production of pulp, and wood is by far the most widely used raw material. Wood, however, differs from northern to southern climates, necessitating different processes for making satisfactory pulp. In some parts of the world where wood is not readily available, other fiber sources are utilized, such as bagasse (sugar cane), bamboo, esparto and hemp. Faster growing sources of pulp like kenaf are also being developed. There are four types of pulping processes: *mechanical, chemical, semi-chemical* and *thermo-mechanical.*

Mechanical pulping produces groundwood pulp. Cleaned and peeled logs are ground against a revolving grindstone, or wood chips are passed between two steel discs or a refiner until they are reduced to fiber. Groundwood pulp is economical since all the wood is used. It does, however, contain impurities which can cause discoloration and weakening of the paper. Its main uses are for newsprint, and as part of the pulp in magazine papers where it contributes bulk, opacity and compressibility.

Chemical pulping removes most of the impurities such as lignin, resins, gums and other undesirable components of the wood so that the pulp is mainly cellulose fiber. Papers made from this pulp are much more permanent (less yellowing or fading) than groundwood paper. Chemical pulping is done by cooking wood chips with chemicals in batch or continuous digesters. There are two main types of chemical wood pulp: *sulfite* and *sulfate.* Sulfite pulp is made by cooking chips of coniferous woods like spruce, pine and hemlock in a liquor made from lime and sulfurous acid. Sulfate pulp, also known as *kraft,* is produced by cooking broadleaf or coniferous woods with caustic soda and sodium sulfide. Since sulfate pulp uses a wider variety of woods and produces a stronger paper, it is used more widely than sulfite pulp.

Semi-chemical pulping combines chemical with mechanical pulping to produce a pulp with higher yield yet somewhat similar properties to chemical pulp. It is usually used as a blend with chemical pulp to impart stiffness and good formation.

Thermo-mechanical pulping (TMP) is a form of mechanical pulping involving the hot pressurized refining of wood chips. By using wood chips and a refiner and steam pressurizing the chip feeder and refiner, pulp is produced with a yield of more than 90 percent from the wood. The strength, particularly tear, is improved considerably compared to groundwood and the pulp is strong enough to be used for newsprint without the addition of chemical pulp.

BLEACHING

The bleaching process is done on pulp to obtain a higher brightness in papers. While pure cellulose is white in color, the presence of impurities and coloring matter gives the pulp a brownish color, as in grocery bags, which are made from unbleached kraft pulp. Chemical pulps have traditionally been bleached in multiple-stage processes (3–7 stages) with chemicals like elemental chlorine, and/or sodium hypochlorite. Because of increasing environmental concerns, bleaching processes have moved away from elemental chlorine, substituting chemicals such as chlorine dioxide, hydrogen peroxide, oxygen, and ozone. These bleaching processes are referred to as elemental chlorine free (ECF).

REFINING

Refining is an important step in papermaking since many characteristics of the paper are largely determined by the treatment of pulp in this operation. Refining is most commonly done in a continuous process. The pulp is passed between a rotating and a stationary set of steel bars which cause cutting and fibulation of the fibers. The treatment is controlled to produce the desired strength and other qualities in the finished paper.

ADDITIVES FOR SPECIAL PROPERTIES

In addition to the refining, certain materials are added to impart other characteristics which make the product more suitable for its intended use. In the now predominant alkaline papermaking process, calcium carbonate is used as a filler to improve opacity and brightness. Dyes and pigments are also added to control color and shade. Sizing, or resistance to water and ink penetration, is imparted by several types of synthetic materials commonly known by their abbreviations as AKD or ASA. In the older, traditional acid-based papermaking process, fillers such as clay are used. Titanium dioxide is added for opacity and brightness. Rosin-alum based internal sizing provides resistance to ink and other liquid penetration.

The combination of fiber, size, fillers and additives is known as *stock*. External sizing, most commonly a starch solution, is added to the surface of the sheet near the end of the paper machine. It acts as a glue to tie down the loose fibers on the surface and close up the pores so the paper can receive ink and toner satisfactorily.

MAKING PAPER

The modern paper machine is extremely complex but consists essentially of three principal units: (1) the forming section, known as the wet end, (2) the press section, where water is removed by pressing the wet paper between rolls and felts, and (3) the drying section, where the moisture content is reduced to the desired level.

The forming section The stock is diluted with process water to less than 1% solids and pumped to a distribution unit or headbox. The headbox spreads the flow to the width of the machine and discharges it through an orifice onto a finely woven endless wire belt. The water is drained through the wire by gravity and suction, leaving the stock on the surface. The fibers tend to align themselves in the direction the machine is traveling.

There is a wide variety of forming sections used to produce paper. Some machines, especially for offset newsprint, are equipped with two wires and are known as twin-wire machines. Twin-wire formers distribute the fines and filler in the z-direction (the paper thickness) producing a sheet that prints cleaner. Other forming sections incorporate a dandy roll, a cylindrical frame covered with wire mesh, located on top of the wire to distribute the fibers and improve formation. In some cases, the surface of the roll contains lettering or a design to produce a watermarked paper.

The press section The paper web leaves the wire still containing 75–85 percent water, and this is reduced to 65% in the press section. This operation is performed in a series of presses, each consisting of two rolls, and the sheet passes through the nip between these rolls supported on a felt belt made mainly of synthetic materials. Removal of water by pressing is more economical than by drying; the presses compact the sheet and level the surface. The location of the press felts (top versus bottom) play a large role in the fines and filler distribution in the sheet.

The drying section Machine drying takes place after the paper leaves the presses and enters the drying sections. In the drying section the sheet is dried to the final moisture content (~5%). The driers are steam-heated cast iron drums, four to six feet in diameter, polished on the outside surface. The drums are generally arranged in two tiers with as many as thirty tiers of driers on some of the larger installations. On paper machines the sheet is held tightly against the driers by a heavy felt usually made of synthetic materials. About two pounds of water are evaporated for each pound of paper produced.

DRY END OF A
PAPER MACHINE

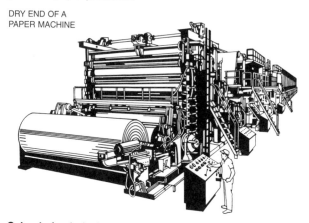

Calendering is the last operation on the paper machine before the paper is wound on reels. Machine calenders are stacks of vertical cast steel rolls that have polished ground surfaces. The paper enters the stack at the top and is compacted and smoothed progressively as it travels down the stack. Calendered papers are known as *machine-finished papers*.

Supercalendering is a subsequent operation. Supercalenders are also arranged vertically, but the rolls are alternately steel and cotton. Supercalenders are used for both coated and uncoated papers.

Supercalendered Papers

Supercalendered papers are used extensively in Europe for printing gravure but were not successful for web offset printing in the U.S. until a special grade called Publication Quality Supercalendered papers (PQSC) was developed in the early 1980s. More than 2,000,000 tons of PQSC papers were used in 1990 and the amount increases annually.

Alkaline Papermaking

Alkaline papermaking was the trend in Europe for years before it became widespread in the U.S. It is now the method by which the majority of fine papers (uncoated wood-free papers) are made. Through this process, calcium carbonate replaces both clay as filler and titanium dioxide as a white pigment. A synthetic compound replaces the rosin-alum mixture for the internal size. Since calcium carbonate is much less costly than titanium dioxide, alkaline papermaking is significantly more economical. Fine papers made by the alkaline process have a longer life. Alkaline paper has improved bulk, brightness and opacity and has better print characteristics. The composition of these papers is covered by the international standards, ISO-9706-T46-SC-10, Paper for documents — Specifications for permanence — 1992.

Recycled Papers

In response to growing demand for recycled paper products, paper recycling is increasing worldwide. In the U.S. recycled newsprint has been used for years, but recycled papers for business and commercial printing only came into moderate use during the late 1980s. Driven by concerns over landfill closures and the need to reduce municipal solid waste, shipments of recycled fiber content printing and writing grades are increasing rapidly. While a small portion of these grades have performance or appearance characteristics not equal to their virgin counterparts, the vast majority exhibit no noticeable differences. Further growth in the production of recycled content grades will require standardized government definitions, some quality wastepaper source separation and collection programs at the local level and possibly the installation of cost effective on-site deinking plants.

Coated Papers

The great popularity of reproducing black-and-white and color photographs brought about the development of coated papers.

These grades reproduce much finer halftone screens with sharper definition, improved density and greater color fidelity than can be reproduced on uncoated papers. Coated paper finishes range from dull to very glossy, have a greater affinity for printing inks, greater smoothness, higher opacity and better ink holdout than uncoated papers.

Coated printing papers are available coated one side only (C1S) for labels, packaging and covers or coated two sides (C2S) for book, publication and commercial printing. Papers coated on the machine are called *machine coated;* those coated on independent coaters are *off-machine coated.*

Coatings consist of suspensions of pigments in suitable binders. They are applied by rolls, air knives or trailing blades. Blade coatings have become very popular since they are smoother, and it is possible to apply lower coat weights which are necessary for lighter weight publication grades. *Cast coated* papers are produced off-machine by pressing newly coated paper against a highly polished chromium-plated drum.

Paper Finishes

Finish is a complex paper property related to its smoothness. Paper can be used as it comes off the driers of a paper machine, or it can be machine calendered and later supercalendered. Uncalendered, machine calendered and supercalendered papers vary greatly in smoothness.

The usual finishes of uncoated book papers are, in order of increasing smoothness: *antique, eggshell, vellum, smooth* and *lustre*. These finishes are classed together because all can be produced on the machine. Additional smoothness is obtained with supercalendering. Coating, of course, further improves the finish and smoothness.

Some finishes are embossed on the paper after it leaves the machine. These are produced by a rotary embosser, a machine similar to a mangle, with the paper passing through it dry and under pressure. Commonly used embossing patterns are linen, tweed and pebble.

Top (Felt) Side and Wire Side

Paper is considered a two-sided material. Each side has different characteristics due to the way paper is made. The side directly in contact with the wire of the paper machine is called the *wire side;* the other is the top or *felt side*. The felt side usually has a closer formation with less grain and better crossing of the fibers. The wire side, however, has less fines on the surface and usually gives less trouble in the depositing of loose paper dust, fiber picking or lint on the blanket of an offset press.

The paper produced on twin-wire machines is less two-sided than paper produced on single-wire machines. Both sides are more similar to the wire side in that they have fewer fines and cause less problems with lint, etc. on offset press blankets. This paper is used extensively for web offset printing.

PAPER CHARACTERISTICS

Grain is an important factor for both printing and binding. It refers to the position of the fibers. During papermaking most fibers are oriented with their length parallel to that of the paper machine *(machine direction)* and their width running across the machine *(cross direction).*

Grain affects paper in the following ways, and these facts need to be considered in the proper use of paper: (1) paper folds smoothly *with* the grain direction and roughens or cracks when folding cross-grain. (2) paper is stiffer in the grain direction, and (3) paper expands or contracts more in the cross direction when exposed to moisture changes.

TEAR AND FOLD TESTS
Paper tears straighter with grain

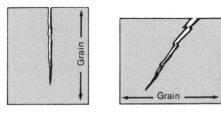

Paper folds more easily with grain

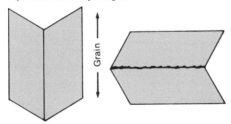

In books and catalogs, grain direction should be parallel with the binding edge. If it is perpendicular with the binding edge, the pages turn less easily and do not lie flat. Paper for sheetfed offset is usually grain long. Moisture changes affect the shorter dimension and register problems are reduced. A second reason for grain long is that the sidewise size of printed images cannot be changed without cutting or changing plates, whereas changing the size of the printed image around the cylinder (short dimension of the sheet) can be accomplished by changing the packing under the plate and blanket.

Basis weight identifies printing papers, with few exceptions. In the U.S. it is the weight in pounds of a ream (500 sheets) in the *basic size* for that grade. In the metric system it is the weight of one square meter of paper and is expressed in grams per square meter or g/m^2. Basis 70 means that 500 sheets 25 x 38 inches of book paper weigh 70 pounds. This is equivalent to 104 g/m^2 in the metric system.

In the U.S. system, the basic size is not the same for all grades: It is 25 x 38 for book papers (coated, text, offset, opaque, etc.); 17 x 22 for writing papers (bond, ledger); 20 x 26 for cover papers (coated and uncoated); 25½ x 30½ for index bristol, 22½ x 28½ or 22½ x 35 for mill bristol and postcard; 24 x 36 for tag and newsprint.

Paper is commonly identified by ream weight: 20-pound bond, 70-pound coated, etc. However, paper is usually listed in size-and-weight tables and price lists on a thousand sheet basis: 25 x 38 — 140M for a 70-pound book paper, the "M" meaning 1,000 sheets 25 x 38 weigh 140 pounds. Each grade is made in many standard sizes other than the basic size, and in many weights. For example, book papers are made in weights from 50-pound to 100-pound in 10-pound increments.

Metric system In the metric system which is used by most countries other than the U.S., basis weight or substance is referred to as *grammage* and is expressed as weight per unit area, or grams per square meter (g/m^2). Such measurements are independent of paper size. There is, however, a basic size of paper in the metric system known as the A series, and while its area is a square meter it is not a meter square. The series consists of sheet sizes in which the area from one size to the next varies by a factor of 2 or ½. Following is a table of the A sizes.

A number	Size (Millimeters)	(Inches)	Area (Square Meters)
A0	841 x 1189	33.1 x 46.8	1.0
A1	594 x 841	23.4 x 33.1	0.5
A2	420 x 594	16.5 x 23.4	0.25
A3	297 x 420	11.7 x 16.5	0.125
A4	210 x 297	8.3 x 11.7	0.063
A5	148 x 210	5.8 x 8.3	0.031

This is part of the basic A series. There are others such as B sizes which are intermediate between the A sizes, RA and SRA stock sizes from which A sizes can be cut, and C series for envelopes and folders suitable for stationery in the A sizes. The standard size in the metric system which corresponds closely to the U.S. 8½″ x 11″ size is A4.

The following table will help paper users and producers to convert basis weights of paper from the U.S. system to the metric system and vice versa.

| Trade Size | Conversion Factor | |
	Metric to U.S. (g/m^2 to lbs.)	U.S. to Metric (lbs. to g/m^2)
17 x 22 — Bond	0.266	3.760
20 x 26 — Cover	0.370	2.704
24 x 36 — Newsprint or Tag	0.614	1.627
25 x 38 — Offset/Book/Text	0.675	1.480
1000 ft.^{2}	0.205	4.831

A simple way to convert approximately the basis weight for 25 x 38 paper to g/m^2 is to multiply it by 1½ or 3/2. 40 lb. book paper is roughly 60 g/m^2 paper. According to the table it is 40 x 1.48 = 59.2 g/m^2. Conversely to convert g/m^2 to lbs./ream 25 x 38 paper, multiply by 2/3. 45 g/m^2 paper is approximately 30 lb. paper. According to the table it is 45 x 0.675 = 30.8.

Thickness is often referred to as caliper and is measured in mils or thousandths of an inch. In book manufacturing, the **bulk** of the paper determines the thickness of the book so it is often expressed in different terms than the thickness or caliper of the sheet. Bulk for book papers is expressed as the number of pages per inch (ppi) for a given basis weight. For example, the bulking range for a 50-pound book paper can be from 310 to 800 ppi.

Strength of paper is more dependent on the nature of its fiber than its thickness. High bursting strength is achieved by closely intermingling long pulp fibers during the forming of the sheet on the paper machine wire. Some papers, paper bags for example, need high tearing-resistance. Fibers are long, and tear resistance in the cross-machine direction is always higher than tear in the machine direction. This is so because the greatest number of fibers lie *across* the path of the crossmachine tear.

Papers which are subjected to considerable tension in use, such as those printed on web presses, should have a high tensile strength as well as high tear strength.

Stretch is the amount of distortion paper undergoes under tensile strain. Stretch is generally much greater in the cross direction than in the machine direction.

RUNNABILITY AND PRINT QUALITY

Two important factors that affect the printing of papers by any process are runnability and print quality. Runnability affects the ability to get the paper through the press and failures in runnability can cause expensive downtime. Print quality factors affect the appearance of the printed image on the paper.

Runnability

This is more of a problem in offset than in letterpress or gravure because of the overall contact of the paper with the blanket during impression, and the use of water and tacky inks. The following paper properties can affect runnability:

Flatness means freedom from buckles, puckers, wave and curl, especially important in offset.

Trimming means sheets should be square, accurate in size.

Dirt is loose material from all manufacturing sources, such as slitter and trimmer dust, lint, starch, anti-set-off spray, loose coating pigments or fillers, or loosely bonded fibers on the surface, and is especially troublesome in offset. The material collects on the blanket and causes spots in printing.

Equilibrium moisture content (ERH) means the paper should be in ERH balance with the pressroom Relative Humidity (RH). An increase in RH can cause *wavy edges*; the edges absorb moisture while the rest of the pile remains unchanged. *Tight edges*, in which the edges lose moisture and contract, are caused when the RH of the pressroom is lower than the paper. Both wavy and tight edges can cause wrinkles and/or misregister in printing, especially in offset.

IMPROPERLY CONDITIONED PAPER

Wavy edges   Tight edges

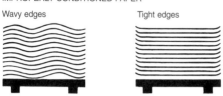

Curl is a moisture-related problem, with the exception of *tail-end hook* and *roll curl*. It is due to stresses in the paper caused by its two-sidedness and differences in the moisture characteristics of the two sides. It is especially severe on coated one-side papers, such as label paper, which is counteracted by special treatments on the back (uncoated) side of the paper. *Tail-end hook* is mechanical curl parallel to the grain caused by heavy solids of ink near the back edge of the sheet. *Roll curl* is mechanical curl, also called *roll set* that is greater near the core of the roll or reel.

Adequate pick resistance can be a problem. Weak paper surfaces tend to pick, blister, delaminate or split when tacky ink is transferred from the plate or blanket to the paper. This is more of a problem in offset than in letterpress.

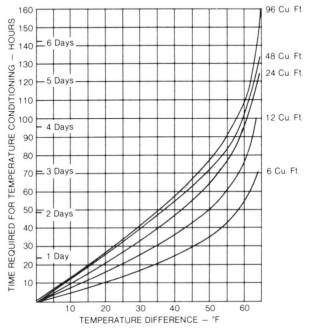

TEMPERATURE CONDITIONING CHART FOR PAPER
Courtesy of Graphic Arts Technical Foundation, Inc.

Water resistance is required since lithographic papers with soluble or water-sensitive coatings tend to pile on the blanket necessitating frequent stops for washups. The piling can occur in the image or non-image areas. Each is caused by different paper properties, and the exact cause is not known for sure.

Paper-ink affinity is important because paper surfaces can affect ink drying, chalking, rub-off, set-off, and ink and varnish holdout. *Ink absorbency* and *holdout* are related. If paper absorbs too much ink (as in newsprint) the images appear weak, desaturated and flat (no gloss). If absorbency is low (as in coated paper) the ink sets near the surface and dries with a reasonable gloss. This is *holdout*. If holdout is too high it can cause *set-off* (transfer to the back of adjacent sheet) in the paper pile.

Sheets with heavy ink coverage may stick together or "block" in the stack.

Web offset print defects are of three types: blistering, delamination and fiber puffing. **Blistering** usually occurs in areas with heavy ink coverage and is caused by moisture in the coated paper web that suddenly vaporizes into steam and ruptures the paper structure internally. **Delamination** appears similar to blistering but can occur in coated or uncoated paper. It is due to the tack of the ink causing the printed paper to follow one of the blankets on a blanket-to-blanket press after passing through the nip, and delaminate (or pull apart) as the paper tension snaps the paper off the blanket. **Fiber puffing** appears as a roughened surface on groundwood-based coated paper due to the swelling of the groundwood fibers, usually because of excessively high temperatures in the dryer.

Mechanical condition refers to the physical condition of the paper sheets or rolls. The paper should be free of holes, wrinkles, torn sheets, scraps, turned-over corners, stuck spots or edges and foreign matter. Paper rolls should be evenly and tightly wound with smooth, even edges and a minimum of splices.

Print Quality

The appearance characteristics of the printed image can be affected by the following paper properties. These are especially important when the same job is being printed in more than one plant or when a job is being reprinted.

Color is important as it affects the color reproduction of lighter tints especially. Paper colors vary with advertising fads from cool to warm shades. Type is more easily read against a soft (yellowish) white, while process colors reproduce most accurately on neutral white paper.

Brightness affects the contrast, brilliance, snap or sparkle of the printed subject. Artificial brighteners, like fluorescent additives, can affect color reproduction since most are not neutral in color and have excess blue reflectance.

Paper color reflectance

Paper	% Reflectance		
	Red	Green	Blue
Paper A	80	80	80
Paper B	80	80	85
Paper C	85	85	80

Paper A is the whitest; paper B is blue-white and is the brightest; and paper C is yellow-white and reflects the most light. The most suitable paper for color reproduction is paper A as it is neutral and all colors printed on it will be in color and gray balance. Colors printed on paper B will appear bluish; colors printed on paper C will appear yellowish or reddish.

Opacity relates to the *show-through* of the printed image from the opposite side of the sheet or the sheet under it. Higher opacity is achieved by increasing mineral filler content or caliper of the paper.

Smoothness is a very important property for letterpress and gravure but has little effect on offset. Smooth surfaces have irregularities of the order of 0.005″ to 0.010″ apart. They cannot be seen by the naked eye but can be detected by a magnifying glass and low angle illumination. As smoothness decreases, solids and halftones get sandy and rough in appearance but type is not affected much.

Paper gloss is the shiny and lustrous appearance of the paper due to the addition of coating and calendering in manufacture.

Printed gloss is related to ink gloss and holdout as well as paper gloss.

Refractiveness relates to light absorption in the surface of the paper, causing halftones to appear darker than they should.

Web offset paper has its grain direction paralleling the direction of web travel. Rolls must be properly wound, protected, stored on end and have good tensile strength to minimize tearing or breaking of the web on press. Paper should be of uniform caliper (thickness) and free from holes, scum spots, slitter dust, fiber picking and lint. It should have a minimum of contraction and expansion, contain a minimum number of splices and have sound cores for winding and delivery. Most paper mills use bar codes for identifying roll position on the log and recording other pertinent information about the paper.

PAPER TESTING AND EVALUATION FOR PRINTABILITY

Printing papers are tested for a number of properties, namely basis weight, brightness, caliper, gloss, oil absorption, opacity, porosity, smoothness, stiffness, tear and tensile strength. While all of these tests are valuable for mill quality control and product uniformity, they are generally useless for predicting the printing characteristics of the paper. Much printability testing has been done, and many test instruments have been designed and built.

But the most reliable testing, for lithography especially, still is done on a production printing press. Most letterpress testing for smoothness, ink receptivity and coverage can be done on special test proof presses.

Printability test instruments can be used to make reasonably accurate predictions of picking, ink coverage, receptivity, and, with newer models, trapping has been predicted with some degree of success. There are special printability testers for gravure. For lithography, however, printability testers are far from having the reliability desired, mainly because the effects of dampening have been difficult to simulate. Bench or proof press tests that correlate with press performance have not been successful, so most printability testing is done on offset presses.

PAPER GRADES

Paper may be defined in terms of its use. Each grade serves a purpose, usually suggested by its grade name. Some of the most common classifications of printing papers are bond, coated, text, cover, book, offset, index, label, tag and newsprint. The size shown in parentheses is the basic size for that grade.

Bond (17 x 22) papers are commonly used for letters and business forms. They have surfaces which accept ink readily from a pen or typewriter and can be easily erased. Most letterheads and business forms are a standard 8½" x 11" size.

Coated (25 x 38) papers are used when high printing quality is desired because of its greater surface smoothness and uniform ink receptivity. There are many kinds: cast coated, gloss coated, dull coated, machine coated, coated one- and two-sides, etc.

Text (25 x 38) papers are noted for their interesting textures and attractive colors. They enjoy frequent use for announcements, booklets and brochures. Most text papers are treated with a sizing to make them more resistant to water penetration and easier to print by offset lithography.

Book (25 x 38) papers are used for trade and textbooks as well as general printing. They are less expensive than text papers, and are made in antique or smooth finishes. Book papers have a wider range of weights and bulk than text papers so it is possible to secure almost any desired bulking.

Offset (25 x 38) papers are similar to the coated and uncoated book paper used for letterpress printing except that sizing is added to resist the slight moisture present in offset printing, and the surface is treated to resist picking.

Cover (20 x 26) papers complement coated and text papers in heavier weights and matching colors for use as covers on booklets, etc. Papers are also made for cover purposes only. Many special surface textures are available, with finishes ranging from antique to smooth. Special characteristics of cover pages include dimensional stability, durability, uniform printing surface, good scoring, folding, embossing and die-cutting qualities. It is a useful rule of thumb that cover stock of the same basis weight as text paper has about twice the thickness.

Index (22½ x 35 and 25½ x 30½) papers have two outstanding characteristics — stiffness and receptivity to writing ink. Index is commonly used whenever an inexpensive stiff paper is required. It is available in both smooth and vellum finish.

Tag (24 x 36) is a utility sheet ranging in weight from 100 to 250 pounds for manufacturing tags. It may be made from sulfite, sulfate or mechanical pulp, and various types of waste papers. Tag board is sometimes tinted and colored on one or both sides. Tag stock has good bending or folding qualities, suitable bursting and tensile strength, good tearing and water resistance, and a surface adaptable to printing, stamping or writing.

Bristol (22½ x 28½) is one of the board grades, with a softer surface than index or tag, making it ideal for high-speed folding, embossing or stamping. It is an economical substitute for cotton fiber stocks, is very receptive to ink and has good snap and resilience.

Newsprint (24 x 36) is used in printing newspapers. Furnish is chiefly groundwood pulp, with some chemical pulp. It is made in basis weights from 28 to 35 pounds, with 30-pound used most extensively.

Lightweight papers such as manifold, onionskin and Bible paper are specialty grades that have been produced for years. Recently, increasing mailing costs have fostered the development and use of lighter weight newsprint and magazine papers. Newsprint as light as 22 pound ($36g/m^2$) has been produced; 32 pound ($47g/m^2$) coated magazine paper is used regularly in web offset and 28 pound ($41g/m^2$) paper in gravure.

CUTTING CHARTS

Informed paper buyers always try to use standard paper sizes which can be used without waste. Odd-size pages can be wasteful and costly if the quantity is not large enough, or if there is not enough time to order a special making-size of paper.

This chart shows the number of pages to a standard paper size for several page sizes in use today. The paper size includes trim top, bottom and side, but not bleed.

Trimmed Page Size	Number of Printed Pages	Number From Sheet	Standard Paper Size
4 x 9	4	12	25 x 38
	8	12	38 x 50
	12	4	25 x 38
	16	6	38 x 50
	24	2	25 x 38
4¼ x 5⅜	4	32	35 x 45
	8	16	35 x 45
	16	8	35 x 45
	32	4	35 x 45
4½ x 6	4	16	25 x 38
	8	8	25 x 38
	16	4	25 x 38
	32	2	25 x 38
5½ x 8½	4	16	35 x 45
	8	8	35 x 45
	16	4	35 x 45
	32	2	35 x 45
6 x 9	4	8	25 x 38
	8	4	25 x 38
	16	2	25 x 38
	32	2	38 x 50
8½ x 11	4	4	23 x 35
	8	2	23 x 35
	16	2	35 x 45
9 x 12	4	4	25 x 38
	8	2	25 x 38
	16	2	38 x 50

EQUIVALENT WEIGHTS

In reams of 500 sheets, basis weights in bold type.

Grade of Paper	BOOK 25 x 38	BOND 17 x 22	COVER 20 x 26	BRISTOL 22½ x 28½	INDEX 25½ x 30½	TAG 24 x 36	METRIC (g/m²)
BOOK	**30**	12	16	20	25	27	44
	40	16	22	27	33	36	59
	45	18	25	30	37	41	67
	50	20	27	34	41	45	74
	60	24	33	40	49	55	89
	70	28	38	47	57	64	104
	80	31	44	54	65	73	118
	90	35	49	60	74	82	133
	100	39	55	67	82	91	148
	120	47	66	80	98	109	178
BOND	33	**13**	18	22	27	30	49
	41	**16**	22	27	33	37	61
	51	**20**	28	34	42	46	75
	61	**24**	33	41	50	56	90
	71	**28**	39	48	58	64	105
	81	**32**	45	55	67	74	120
	91	**36**	50	62	75	83	135
	102	**40**	56	69	83	93	158
COVER	91	36	**50**	62	75	82	135
	110	43	**60**	74	90	100	163
	119	47	**65**	80	97	108	176
	146	58	**80**	99	120	134	216
	164	65	**90**	111	135	149	243
	183	72	**100**	124	150	166	271
BRISTOL	100	39	54	**67**	81	91	148
	120	47	65	**80**	98	109	178
	148	58	81	**100**	121	135	219
	176	70	97	**120**	146	162	261
	207	82	114	**140**	170	189	306
	237	93	130	**160**	194	216	351
INDEX	110	43	60	74	**90**	100	163
	135	53	74	91	**110**	122	203
	170	67	93	115	**140**	156	252
	208	82	114	140	**170**	189	328
TAG	110	43	60	74	90	**100**	163
	137	54	75	93	113	**125**	203
	165	65	90	111	135	**150**	244
	192	76	105	130	158	**175**	284
	220	87	120	148	180	**200**	326
	275	109	151	186	225	**250**	407

Comparative Weights of Book Papers per 1,000 Sheets

Basis	50	60	70	80	100	120
8½ x 11	9.8	11.8	13.8	15.7	19.7	23.6
17½ x 22½	41	50	58	66	83	99
19 x 25	50	60	70	80	100	120
23 x 29	70	84	98	112	140	169
23 x 35	85	102	119	136	169	203
24 x 36	90	110	128	146	182	218
25 x 38	100	120	140	160	200	240
35 x 45	166	198	232	266	332	398
36 x 48	182	218	254	292	364	436
38 x 50	200	240	280	320	400	480
*metric (g/m^2)	74	89	104	118	148	178

*Metric equivalent of basis weight.

ENVELOPE STYLES

A. Commercial envelopes are used for business correspondence, either surface or airmail. Made from bond and kraft papers in all standard sizes.

B. Window envelopes are used primarily for statements, dividends and invoices. The window saves time and prevents an element of error by eliminating typing of an extra address. Window envelopes are made in all sizes and styles, from many types of paper.

C. Self-Sealing envelopes have latex adhesive on upper and lower flaps that seal instantly without moisture when flaps come together. These envelopes are a time saver in handling.

D. Booklet, Open-Side envelopes are ideal for direct mail and house organs. A concealed seam lends itself to overall printing in front and back.

E. Baronial envelopes are a more formal open-side envelope with a deep, pointed flap. They are often used for invitations, greeting cards, announcements, etc.

F. Bankers Flap and Wallet Flap envelopes handle unusually bulky correspondence. They can be crammed with correspondence and will carry material safely. Reserve strength is far in excess of everyday commercial envelopes.

G. Clasp and String-and-Button envelopes are sturdy and widely used for mailing bulky papers. Metal clasps are smooth and burrless. String and button keep contents under tension and better protected in the mail. Both types may be opened and closed many times.

H. Open-End envelopes are used for mailing catalogs, reports, booklets and magazines. Wide seams and heavy gummed flaps ensure maximum protection under rough handling conditions.

I. Expansion envelopes are used for bulky correspondence and for package and rack sales.

ENVELOPE STYLES

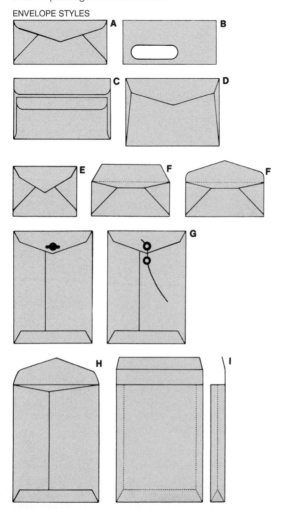

Graphic Arts Terms

Bas
trim
grain
stripping
electr
widow
camper
spine
case
PDF
anilox inking
poir
UV
WYSIWYG

INTERNATIONAL PAPER

To list all terms connected with the graphic arts would fill a book. In this section the most common terms, abbreviations and acronyms used in graphic arts production and the new technologies are defined. In lieu of an index many terms have references to pages where the terms are used or discussed.

AA Abbreviation for *author's alterations.*

absorption *In paper,* the property which causes it to take up liquids or vapors in contact with it. *In optics,* the partial suppression of light through a transparent or translucent material.

accordion fold *In binding,* a term used for two or more parallel folds which open like an accordion. *(page 175)*

Acrobat Adobe software that embodies the PDF format. *(page 105)*

A/D converter Device or software to convert an analog signal to a digital signal. *See digitizer.*

additive primaries *In color reproduction,* red, green and blue (RGB). When lights of these colors are added together, they produce the sensation of white light. *(page 61)*

against the grain Folding or feeding paper at right angles to the grain direction of the paper. Also called crossgrain. *(page 187)*

airbrush *In artwork,* a small pressure gun shaped like a pencil that sprays watercolor pigment. Used to correct and obtain tone or graduated tone effects. *In platemaking,* used with an abrasive-like pumice to remove spots or other unwanted areas. *In electronic imaging,* a retouching technique.

alkaline paper Paper made with a synthetic alkaline size and an alkaline filler like calcium carbonate which gives the paper over four times the life (200 years) of acid-sized papers (40-50 years). *(page 186)*

AM (Amplitude Modulation) Halftone screening, as opposed to FM screening, has dots of variable size with equal spacing between dot centers. *See halftone. (page 71)*

analog color proof Off-press color proof made from separation films. *(page 72)*

anilox inking *In flexography,* two-roll inking system with a smooth fountain roll that transfers inks to an etched metal or ceramic-coated metal roll with cells of fixed size and depth that transfer the ink to the plate. Also used in *keyless offset.* *(page 141)*

anti-halation backing *In photography,* coating applied to back of film to prevent halation.

anti-offset or set-off spray *In printing,* dry spray of finely powdered starch used on press to prevent wet ink from transferring from the top of one sheet to the bottom of the next sheet.

antique finish A term describing the surface, usually on book and cover papers, that has a natural rough finish. *(page 187)*

aperture *In photography,* lens opening or lens stop expressed as an f/no. such as f/22.

apochromatic *In photography,* color-corrected lenses which focus the three colors, blue, green and red, in the same plane.

APR (Automatic Picture Replacement) The replacement of a low resolution image by a high resolution image. *(page 99)*

argon laser A very strong blue laser that peaks at 470 nanometers.

art All illustration copy used in preparing a job for printing. *(page 56)*

ascender That part of a lower-case letter which rises above the main body, as in "b". *(page 36)*

ASCII (American Standard Code for Information Interchange) A standard means of representing text as numerical data.

automatic processor *In photography,* a machine to automatically develop, fix, wash and dry exposed photographic film. *(page 65) In platemaking,* a machine to develop, rinse, gum and dry printing plates. *(page 119)*

backbone The back of a bound book connecting the two covers; also called spine. *(page 177)*

backing up Printing the reverse side of a sheet already printed on one side.

bad break *In composition,* starting a page or ending a paragraph with a single word, or *widow.*

basic size In inches, 25 x 38 for book papers, 20 x 26 for cover papers, 22½ x 28½ or 22½ x 35 for bristols, 25½ x 30½ for index. *(page 189)*

basis weight The weight in pounds of a ream (500 sheets) of paper cut to a given standard size for that grade; e.g., 500 sheets 25″ x 38″ of 50-lb. book paper weigh 50 pounds. *(page 188)*

bearers *In presses,* the flat surfaces or rings at the ends of cylinders that come in contact with each other during printing and serve as a basis for determining packing thickness.

beta site A test site for computer software or systems.

bezier curve The description of a character, symbol or graphic by its outline used by drawing programs to define shapes. *(page 101)*

bimetal plate *In lithography,* a plate used for long runs in which the printing image base is usually copper and the non-printing area is aluminum, stainless steel or chromium. *(page 121)*

bit *In computers,* the basic unit of digital information; contraction of BInary digiT. *(page 47)*

bitmap *In computer imaging,* the electronic representation of a page, indicating the position of every possible spot (zero or one). *(page 101)*

black-and-white Originals or reproductions in single color, as distinguished from multicolor. Abbreviation: B/W.

black printer *In color reproduction,* the black plate, made to increase contrast of dark tones and make them neutral. *(pg. 69)*

blanket *In offset printing,* a rubber-surfaced fabric which is clamped around a cylinder, to which the image is transferred from the plate, and from which it is transferred to the paper.

bleed An extra amount of printed image which extends beyond the trim edge of the sheet or page.

blind embossing A design which is stamped without metallic leaf or ink, giving a bas-relief effect. *(page 179)*

blind image *In lithography,* an image that has lost its ink receptivity and fails to print.

blowup An image enlargement.

body *In inkmaking,* a term referring to the viscosity, or consistency, of an ink (e.g., an ink with too much body is stiff). *(page 157)*

body type A type used for the main part or text of a printed piece, as distinguished from the heading.

bold-face type A name given to type that is heavier than the text type with which it is used.

bond paper A grade of writing or printing paper where strength, durability and permanence are essential requirements; used for letterheads, business forms, etc. The basic size is 17″ x 22″. *(page 195)*

book paper A general term for coated and uncoated papers. The basic size is 25″ x 38″. *(page 195)*

bpi Bits per inch

bps Bits per second

break for color *In artwork* and *composition,* to separate the parts to be printed in different colors.

brightness *In photography,* light reflected by the copy. *In paper,* the reflectance or brilliance of the paper. *(page 193)*

brochure A pamphlet bound in booklet form.

bronzing Printing with a sizing ink, then applying bronze powder while still wet to produce a metallic lustre. *(page 162)*

bulk The degree of thickness of paper. *In book printing,* the number of pages per inch for a given basis weight. *(page 190)*

bump exposure *In photography,* an exposure in halftone photography, especially with contact screens, in which the screen is removed for a short time. It increases highlight contrast and drops out the dots in the whites. *(page 67)*

burn *In platemaking,* a common term used for a plate exposure.

byte *In computers,* a unit of digital information, equivalent to one character or 8 to 32 bits. *(page 47)*

CADD (Computer-Aided Drafting or Design) *In graphics,* the production of drawings and plans for architecture and engineering systems. CADD systems are specialized workstations or high-performance personal computers that employ CADD software packages and input devices such as graphic tablets and scanners.

calender rolls A set or stack of horizontal cast-steel rolls with polished ground surfaces at the end of a paper machine. The paper is passed between the rolls to increase the smoothness and gloss of its surface. *(page 185)*

caliper The thickness of paper, usually expressed in thousandths of an inch (mils). *(page 190)*

camera-ready Copy which is ready for photography.

caps and small caps Two sizes of capital letters made in one size of type, commonly used in most roman typefaces.

case *In bookbinding,* the covers of a hardbound book.

cast coated Coated paper dried under pressure against a polished drum to produce a high-gloss enamel finish. *(page 187)*

CCD (Charge Coupled Device) *In digital prepress,* a semiconductor light sensitive electronic device that emits an electrical signal proportional to the amount of light striking it. Used in scanners and video cameras. *(page 93)*

CD-ROM (Compact Disc Read Only Memory) *In digital prepress,* a laser encoded optical storage disc that can store 650 Megabytes to over 1 Gigabyte of data on a disc about the size of a traditional 5-inch floppy disk. *(page 49)*

CEPS (Color Electronic Prepress System) *In digital prepress,* a high-end computer-based system that is used to color correct scanner images and assemble image elements into final pages. They are device dependent systems. *(page 94)*

chalking *In printing,* a term which refers to improper drying of ink. Pigment dusts off because the vehicle has been absorbed too rapidly into the paper.

character generation The production of typographic images using font master data. Generated to screens or output devices.

chemical pulp *In papermaking,* treatment of groundwood chips with chemicals to remove impurities such as lignin, resins and gums. There are two types, sulfite and sulfate. *(page 182)*

chemistry *In photography* and *platemaking,* a term used to describe the composition of processing solutions.

chokes and spreads Overlap of overprinting images to avoid color or white fringes or borders around image detail. Called *trapping* in digital imaging systems.

CIE color spaces These are three dimensional color mapping systems such as CIELab, CIEL*a*b*, and CIELUV which are used to plot the three color attributes, X, Y, Z. These systems are not discussed in this book.

closed loop system *In printing,* a completely automatic control system.

CMY (Cyan, Magenta, Yellow) Subtractive primary colors, each of which is a combination of two additive primary colors (RCB). *(page 62)*

CMYK (Cyan, Magenta, Yellow, Black) The subtractive process colors used in color printing. Black (K) is added to enhance color and contrast. *(page 62)*

coated paper Paper having a surface coating which produces a smooth finish. Surfaces vary from eggshell to glossy. *(page 186)*

coating *In platemaking,* the light-sensitive polymer or mixture applied to a metal plate. *(page 116) In printing,* an emulsion, varnish or lacquer applied over a printed surface to protect it.

cold color *In printing,* a color with a bluish cast.

collate *In binding,* the gathering of sheets and signatures. *(page 175)*

collotype A screenless printing process of the planographic ink-water type in which the plates are coated with bichromated gelatin, exposed to continuous-tone negatives, and printed on lithographic presses with special dampening. *(page 138)*

color balance The correct combination of cyan, magenta and yellow to (1) reproduce a photograph without a color cast, (2) produce a neutral gray, or (3) reproduce the colors in the original scene or object.

color correction Any method such as masking, dot-etching, re-etching and scanning, used to improve color rendition. *(page 69)*

color filter A sheet of dyed glass, gelatin or plastic, or dyed gelatin cemented between glass plates, used in photography to absorb certain colors and transmit others. The filters used for color separation are red, green and blue. (RGB) *(page 62)*

colorimeter An instrument for measuring color the way the eye sees color. *(page 168)*

color keys Off-press overlay color proofs using 3M Color Key® materials.

color proofs *See analog color proof, digital color proof. (pg. 72)*

color separation *In photography,* the process of separating color originals into the primary printing color components in negative or positive form using RGB filters. *(page 68)*

commercial register Color printing on which the misregister allowable is within ± one row of dots.

common impression cylinder press *In flexography, letterpress lithography* and *digital printing,* a press with a number of printing units around a large impression cylinder.

commingled mailing Combined mailing of magazines of the same size to the same address to save costs. *(page 180)*

computer, analog A computer that solves a mathematical problem by using analogs, like voltage or density, of the variables in the problem.

computer, digital A computer that processes information in discrete digital form.

computerized composition An all-inclusive term for the use of computers to automatically perform the functions of hyphenation, justification and page formatting.

computer-to-plate See CTP.

condensed type A narrow or slender typeface.

conductivity A property of fountain solutions that must be controlled along with pH. *(page 137)*

contact print A photographic print made from a negative or positive in contact with sensitized paper, film or printing plate. *(page 66)*

contact screen A halftone screen on film having a dot structure of graded density, used in vacuum contact with the photographic film to produce halftones. *(page 66)*

continuous tone An image which contains gradient tones from black to white. *(page 62)*

contone Abbreviation for continuous tone.

contract proof A color proof representing an agreement between the printer and the customer regarding how the printed product will look. *(page 113)*

contrast The tonal gradation between the highlights, middle tones and shadows in an original or reproduction. *(page 67)*

copy Any furnished material (typewritten manuscript, pictures, artwork, etc.) to be used in the production of printing. *(page 55)*

copy preparation Directions for, and checking of, desired size and other details for illustrations, and the arrangement into proper position of various parts of the page to be photographed or electronically processed for reproduction.

cover paper A term applied to a variety of papers used for the covers of catalogs, brochures, booklets and similar pieces.

crop To eliminate portions of the copy, usually on a photograph or plate, indicated on the original by *cropmarks*. *(page 59)*

cross direction *In paper,* the direction across the grain. Paper is weaker and more sensitive to changes in relative humidity in the cross direction than the grain direction. *(page 187)*

crossmarks *See register marks.*

CTP (Computer To Plate) *In platemaking,* Computer-to-Plate systems or platesetters eliminate the need for having a separate film-to-plate exposure system. *(pages 31, 122)*

curl *In paper,* the distortion of a sheet due to differences in structure or coatings from one side to the other, or to absorption of moisture on an offset press. *(page 191)*

cutoff *In web printing,* the cut or print length. *(page 135)*

cutscore *In die-cutting,* a sharp-edged knife, several thousandths of an inch lower than the cutting rules in a die, made to cut part way into the paper or board for folding purposes. *(page 179)*

cyan Hue of a subtractive primary and a 4-color process ink. It reflects or transmits blue and green light and absorbs red light. *(page 62)*

cylinder gap *In printing presses,* the gap or space in the cylinders of a press where the mechanism for plate (or blanket), clamps and grippers (sheetfed) is housed.

dampeners *In lithography,* cloth-covered, parchment paper or rubber (bare-back) rollers that distribute the dampening solution to the press plate or ink roller. *(page 137)*

dampening system *In lithography,* the mechanism on a press for transferring dampening solution to the plate during printing. *(page 137)*

DCS (Desktop Color Separation) *In digital prepress,* a data file standard defined to assist in making color separations with desktop publishing systems. Using DCS five files are created: four color files, containing the cyan, magenta, yellow and black image data, and a composite color viewfile of the color image. *(page 99)*

deckle *In papermaking,* the width of the wet sheet as it comes off the wire of a paper machine.

deckle edge The untrimmed feathery edges of paper formed where the pulp flows against the deckle.

densitometer *In photography,* a photoelectric instrument which measures the density of photographic images, or of colors. *In printing,* a reflection densitometer is used to measure and control the density of color inks on the substrate. *(page 167)*

density The degree of darkness (light absorption or opacity) of a photographic image. *(page 167)*

descender That part of a lower case letter which extends below the main body, as in "p". *(page 36)*

desktop publishing Process of composing pages using a standard computer, off-the-shelf software, a device independent page description language like PostScript and outputting them on a printer or imagesetter. *(page 95)*

developer *In photography,* the chemical agent and process used to render photographic images visible after exposure to light. *In lithographic platemaking,* the material used to remove the unexposed coating. *(page 65)*

device dependent A characteristic of CEPS. *See CEPS.*

device independent The characteristic of a computer program or system that allows different output devices to image the same file more or less the same. *(page 95)*

diazo *In photography,* a non-silver coating for contact printing. *In offset platemaking,* a light-sensitive coating used on presensitized and wipe-on plates. *(page 120)*

diecutting The process of using sharp steel rules to cut special shapes for labels, boxes and containers, from printed sheets. Diecutting can be done on either flatbed or rotary presses. Rotary diecutting is usually done inline with the printing. *(page 179)*

die-stamping An intaglio process for the production of letter-heads, business cards, etc., printing from lettering or other designs engraved into copper or steel. *(page 143)*

diffusion transfer *In photography* and *platemaking*, a system consisting of a photographic emulsion on which a negative is produced, and a receiver sheet on which a positive of the image is transferred during processing. *(page 130)*

digital color proof A color proof produced from digital data without the need for separation films. *(page 73)*

digital inks *See toners. (page 165)*

digital plates Printing plates that can be exposed by lasers or other high energy sources driven by digital data in a platesetter. *(page 116)*

digital printing Printing by plateless imaging systems that are imaged by digital data from prepress systems. *(page 147)*

digitizer A computer peripheral device that converts an analog signal (images or sound) into a digital signal.

dimensional stability Ability to maintain size; resistance of paper or film to dimensional change with change in moisture content or relative humidity.

direct screen halftone *In color separation,* a halftone negative made by direct exposure from the original on an enlarger or by contact through a halftone screen.

display type *In composition,* type set larger than the text.

dithering *In computer graphics,* a technique for alternating the values of adjacent dots or pixels to create the effect of inter-mediate values. Dithering refers to the technique of making different colors for adjacent dots or pixels to give the illusion of a third color. *(pages 48, 101)*

doctor blade *In gravure,* a knife-edge blade pressed against the engraved printing cylinder which wipes away the excess ink from the non-printing areas. *(page 126)*

DOS (Disk Operating System) *In digital imaging,* a program containing instructions for a computer to read and write data to and from a disk. An operating system (set of programs) that instructs a disk-based computing system to manage resources and operate peripheral equipment.

dot The individual element of a halftone. In AM screening the dots vary in size. In FM screening the dots are very small and usually all the same size. *(pages 47, 71, 72)*

dot gain *In printing,* a defect in which dots print larger than they should, causing darker tones or stronger colors. *(page 138)*

dots per inch (dpi) A measure of the resolution of a screen image or printed page.

download Sending information to another computer or to an output.

draw-down *In inkmaking,* a term used to describe ink chemist's method of roughly determining color shade. A small glob of ink is placed on paper and drawn down with the edge of a putty knife spatula to get a thin film of ink.

drop-out Portions of originals that do not reproduce, especially colored lines or background areas (often on purpose).

drum scanner Uses photo multiplier tubes (PMT) and produces color separations with higher resolution and dynamic range than CCD scanners. *(page 92)*

dryer *In inkmaking,* a substance added to hasten drying.

DTP Acronym for Desktop Publishing.

dummy A preliminary layout showing the position of illustrations and text as they are to appear in the final reproduction. A set of blank pages made up in advance to show the size, shape, form and general style of a piece of printing. *(page 56)*

duotone *In photomechanics,* a term for a two-color halftone reproduction from a one-color photograph. *(page 82)*

duplex paper Paper with a different color or finish on each side.

duplicating film A film for making positives from positives, and negatives from negatives. *In color reproduction,* a special film used for making duplicates of color transparencies. *(page 66)*

DVD (Digital Video or Versatile Disk) A CD-ROM that can store audio, video and computer data at four or more gigabytes per disk. *(page 49)*

dynamic range Density difference between highlights and shadows of scanned subjects. *(page 70)*

electronic dot generation (EDG) *In digital imaging,* a method of producing halftones electronically on scanners and prepress systems. *(page 67)*

electronic printing *In digital printing,* any technology that reproduces pages without the use of traditional ink, water or chemistry or plates. Also known as plateless printing. *(page 24)*

electrophotography Image transfer systems used in copiers to produce images using electrostatic forces and toners. *(page 30)*

electrostatic assist *In gravure,* use of electrostatic forces to help draw ink from gravure cells to reduce *skips* in highlights. *(page 143)*

electrostatic plates Plates for high speed laser printing using zinc oxide or organic photoconductors.

electrotype Duplicate relief plate used for letterpress printing. *(page 125)*

elliptical dot *In halftone photography,* elongated dots which give improved gradation of tones particularly in middle tones and vignettes—also called *chain dots. (page 66)*

em *In composition,* a unit of measurement exactly as wide and high as the point size being set. So named because the letter "M" in early fonts was usually cast on a square body. *(page 42)*

embossed finish Paper with a raised or depressed surface resembling wood, cloth, leather or other pattern. *(page 187)*

embossing Impressing an image in relief to achieve a raised surface; either overprinting or on blank paper (called *blind embossing*). *(page 179)*

EME (electromechanical engraver) *In gravure,* machine used to make gravure printing cylinders. *(page 127)*

emulsion side *In photography,* the side of the film coated with the silver halide emulsion.

en *In composition,* one-half the width of an em. *(page 42)*

enamel A term applied to a coated paper or to a coating material on a paper.

english finish A grade of book paper with a smoother, more uniform surface than machine finish.

EPS (Encapsulated PostScript) *In digital prepress,* a file format used to transfer graphic images within compatible applications. A file containing structured PostScript code, comments and a screen display image. *(page 104)*

etch *In offset lithography,* an acidified gum solution used to desensitize the non-printing areas of the plate; also, an acid solution added to the fountain water to help keep non-printing areas of the plate free from ink.

expanded type A type whose width is greater than normal.

exposure *In photography and platemaking,* the step in photographic or photomechanical processes during which light or other radiant energy produces the image on the photo-sensitive coating.

Fadeometer An instrument used to measure the fading properties of inks and other pigmented coatings.

fake color *In color reproduction,* producing a color illustration by using one image as a key and making the other separations from it manually.

fanout *In printing,* distortion of paper on the press due to waviness in the paper caused by absorption of moisture at the edges of the paper, particularly across the grain.

feeder *In printing presses,* the section that separates the sheets and feeds them in position for printing.

felt side The smoother side of the paper for printing. The top side of the sheet in paper manufacturing. *(page 187)*

file A group of related information, such as text, graphics, page instructions and picture information stored on magnetic disks.

filling in (or filling up) *In letterpress or offset lithography,* a condition where ink fills the area between the halftone dots or plugs up (fills in) the type.

fixing Chemical action following development to convert unexposed silver halide to a water-soluble salt and make the image stable and insensitive to further exposure.

flash exposure *In halftone photography,* the supplementary exposure given to strengthen the dots in the shadow areas of negatives. *(page 67)*

flat *In offset lithography,* the assembly of negatives on goldenrod paper or positives on film, ready for platemaking. *(page 76)* *In photography,* a photograph lacking in contrast.

flatbed scanner A device that scans images in a manner similar to a photocopy machine; the original art is positioned face down on a glass plate. *(page 93)*

flush cover A cover that has been trimmed to the same size as the inside text pages as in this book.

flush left (or right) *In composition,* type set to line up at the left (or right). This page is set flush left *and* right.

flush paragraph A paragraph with no indention.

flying paster *In web printing,* an automatic pasting device that splices a new roll of paper onto an expiring roll, without stopping the press.

FM (Frequency Modulation) screening A means of digital screening. *See stochastic screening.*

focal length *In photography,* the distance from the center of the lens to the image of an object at infinity. At same size, the distance from copy to image is four times the focal length of the lens.

fog *In photography,* silver density in the non-image areas.

folio The page number.

font *In composition,* a complete assortment of letters, numbers, punctuations, etc., of a given size and design. *(page 36)*

form *In offset,* the assembly of pages and other images for printing. *In letterpress,* type and other matter locked in a chase for printing.

form rollers The rollers, either inking or dampening, which directly contact the plate on a printing press.

format The size, style, type page, margins, printing requirements, etc., of a printed piece. *(page 104)*

fountain solution *In lithography,* a solution of water, a natural or synthetic gum and other chemicals used to dampen the plate and keep non-printing areas from accepting ink. *(page 137)*

FPO (For Position Only) *In digital imaging,* typically a low-resolution image positioned in a document to be replaced later with a higher resolution version of the same image. *(page 98)*

free sheet Paper free of mechanical wood pulp.

front end system *In electronic publishing,* the workstation or group of workstations containing the applications software for preparing pages of type and graphics.

"f" stops *In photography,* fixed stops for setting lens apertures.

galley proof A proof of text copy before being made into pages.

gamma A measure of contrast in photographic images.

gapless Plate or blanket cylinders without gaps. *(page 135)*

GATF (Graphic Arts Technical Foundation)

gathering *In binding,* the assembling of folded signatures in proper sequence. *(page 175)*

GCR (Gray Component Replacement) *(page 69)*

gear streaks *In printing,* parallel streaks appearing across the printed sheet at the same interval as gear teeth on the cylinder.

generation Each succeeding stage in reproduction from the original copy.

gigabyte (GB) One billion bytes.

goldenrod paper *In offset lithography,* a specially-coated masking paper of yellow or orange color used by strippers to assemble and position negatives for exposure on plates. *(page 76)*

grain *In papermaking,* the direction in which most fibers lie which corresponds with the direction in which the paper is made on a paper machine. *(page 187)*

grammage A term in the metric system for expressing the basis weight of paper. It is the weight in grams of a square meter of the paper expressed in g/m^2. *(page 189)*

graphical user interface *See GUI.*

gray balance The dot values or densities of cyan, magenta and yellow that produce a neutral gray.

gray level The number of gray values that can be distinguished by a color separation filter — usually 2^8 or 256.

gray scale A strip of standard gray tones, ranging from white to black, placed at the side of original copy during photography to measure tonal range and contrast (gamma) obtained. *(pages 67 and 168)*

gripper edge The leading edge of paper as it passes through a printing press. Also, the front edge of a lithographic or wraparound plate secured to the front clamp of a plate cylinder.

gripper margin Unprintable blank edge of paper on which grippers bear, usually ½″ or less.

grippers *In sheetfed printing presses,* metal fingers that clamp on paper and control its flow as it passes through.

groundwood pulp A mechanically-prepared wood pulp used in the manufacture of newsprint and publication papers. *(page 182)*

GUI (Graphical User Interface) Pronounced "gooey," *in digital imaging,* a technical term for a system that lets users manipulate files by pointing to pictures (icons) with a mouse or other pointing device instead of having to type in key commands. *(page 97)*

gum arabic *In offset lithography,* used in platemaking and on press to protect the non-printing areas of plates.

gumming *In platemaking,* the process of applying a thin coating of gum to the non-printing areas of a lithographic plate.

gutter The blank space or inner margin from printing area to binding. *(page 75)*

hairline register Register within ± ½ row of dots. *(page 58)*

halation *In photography,* a blurred effect, resembling a halo, usually occurring in highlight areas or around bright objects.

halftone The reproduction of continuous-tone images, through a screening process, which converts the image into dots of various sizes and equal spacing between centers (AM screening), or dots of equal size with variable spacing between them (FM screening). *(page 62)*

hard dot Halftone dot with little or no fringe and prints with little or no dot gain or sharpening. *See soft dot.*

hard copy The permanent visual record of the output of a computer or printer on a substrate.

hard proof A proof on paper or other substrate as distinguished from a *soft proof* which is an image on a VDT screen. *(page 72)*

hardware Computer and peripherals as distinguished from *software* which is a program for operating hardware.

head margin The white space above first line on a page.

He/Ne Helium-Neon red laser with wave length of 632 nm.

hickeys *In offset lithography,* spots or imperfections in the printing due to dirt on the press, dried ink skin, paper particles, etc.

high contrast *In photography,* a reproduction with high gamma in which the difference in darkness (density) between neighboring areas is greater than in the original. *(page 68)*

highlight The lightest or whitest parts in a photograph represented in a halftone reproduction by the smallest dots or the absence of dots.

holdout *In printing*, a property of coated paper with low ink absorption which allows ink to set on the surface with high gloss. Papers with too much holdout cause problems with set-off.

HSV Acronym for hue, saturation and value (or brilliance or luminance) — a color space used in some graphic programs.

HTML (HyperText Markup Language) *In imaging for the World Wide Web,* the coding language that is used to create Hypertext documents for use on the World Wide Web. *(page 53)*

hue *In color,* the main attribute of a color which distinguishes it from other colors.

hydrophilic Water receptive.

hydrophobic Water repellent.

hypertext Links to other documents. Words or phrases in the document that are so defined that they can be selected and then cause another document to be retrieved, opened, and displayed. *(page 54)*

image assembly *See stripping.*

imagesetter *In digital imaging,* a generic term that applies to film-output devices for type and graphics. The difference between an imagesetter and a typesetter is in the format of the data that has been converted from discrete-character raster lines to raster data using bitmaps. *(page 110)*

imposetter *In digital imaging,* an imagesetter capable of outputting a film flat with 4, 8 or more pages in imposed position. *(page 111)*

imposition *In image assembly,* the positioning of pages on a signature so that after printing, folding and cutting, all pages will appear in the proper sequence. *(page 74)*

impression cylinder *In printing,* the cylinder on a printing press against which the paper picks up the impression from the inked plate in direct printing, or the blanket in offset printing.

ink fountain *In printing presses,* the device which stores and supplies ink to the inking rollers.

ink-jet printing *In digital printing,* a plateless printing system that produces images directly on paper from digital data using streams of very fine drops of dyes which are controlled by digital signals to produce images on paper. *(page 150)*

ink mist Flying filaments or threads formed by long low-tack inks like newspaper ink. *See long ink.*

Inkometer *In ink testing,* an instrument for measuring the tack of printing inks. *(page 169)*

insert A printed piece prepared for insertion into a publication or another printed piece.

IR Abbreviation for infrared radiation above 700 nm.

italic The style of letters that slant, in distinction from upright, or roman, letters. Used for emphasis within the text. *(page 10)*

jog To align sheets of paper into a compact pile.

justify *In composition,* to space out lines uniformly to line up left and right.

kerning *In typesetting,* subtracting space between two characters, making them closer together. *(page 41)*

key To code copy to a dummy by means of symbols, usually letters. Insertions are sometimes *keyed* in like manner.

keyboard The input device to input information directly into a typesetter, computer, workstation or, as a stand-alone unit, to record it on paper or magnetic tape.

keyline *In artwork,* an outline drawing of finished art to indicate the exact shape, position and size for such elements as halftones, line sketches, etc.

kilobyte (K or kb or KB) 1024 bytes, the most common measure of computer file length.

kiss impression *In printing,* a very light impression, just enough to produce an image on the paper.

kraft A paper or board containing unbleached wood pulp (brown in color) made by the sulfate process. *(page 182)*

lacquer A clear resin/solvent coating, usually glossy, applied to a printed sheet for protection or appearance. *(page 164)*

laid paper Paper with a pattern of parallel lines at equal distances, giving a ribbed effect.

lamination A plastic film bonded by heat and pressure to a printed sheet for protection or appearance.

LAN (Local Area Network) Communication link in a localized area, such as an office, building, complex of buildings or campus, with technology that provides a high-bandwidth, low-cost medium to which many computer nodes can be connected. *(page 110)*

laser (Light Amplification by Stimulated Emission of Radiation) The laser is an intense light beam with very narrow bandwidth used in digital-imaging devices to produce images by electronic impulses from computers or facsimile transmission.

layout The drawing or sketch of a proposed printed piece. *(page 56) In platemaking,* a sheet indicating the settings for a step-and-repeat machine.

leaders *In composition,* rows of dashes or dots to guide the eye across the page. Used in tabular work, programs, tables of contents, etc.

leading (pronounced *ledding) In composition,* the distance between lines of type measured in points. *(page 41)*

LED Acronym for *light emitting diodes* that are used in place of lasers for some output systems.

ledger paper A grade of business paper generally used for keeping records where it is subjected to appreciable wear so it requires a high degree of durability and permanence.

letterspacing The placing of additional space between each letter of a word.

line copy Any copy suitable for reproduction without using a halftone screen.

local area network (LAN) *See LAN.*

logotype (or logo) The name of a company or product in a special design used as a trademark in advertising.

long ink An ink that has good flow on ink rollers of a press. If the ink is too long, it breaks up into filaments on the press, and causes *flying* as on a newspaper press. *(page 157)*

lower case The small letters in type, as distinguished from the capital letters. *(page 36)*

lpi Acronym for lines per inch.

M Abbreviation for Mega, which is commonly used to mean one million. In computer terminology, however, M refers to the number 1,048,576, and is used to specify the amount of storage available on a disk or in memory. A 1 megabyte disk can record 1,048,576 bytes of data. *See megabyte.* Also, abbreviation for quantity of 1,000.

machine coated Paper which is coated one- or two-sides on a paper machine. *(page 186)*

machine direction Same as grain direction in paper. *(pg. 187)*

magenta Hue of a subtractive primary and a four-color process ink. It reflects or transmits blue and red light and absorbs green light. *(page 62)*

magenta screen A dyed contact screen, used for making halftones. *(page 66)*

magnetic storage Any disc, film, tape, drum or core that is used to store digital information.

makeover *In platemaking,* a plate which is remade.

makeready *In printing,* all work done to set up a press for printing. *(page 140)*

mask *In color separation photography,* an intermediate photographic negative or positive used in color correction. *In offset lithography,* opaque material used to protect open or selected areas of a printing plate during exposure.

master A plate for a duplicating machine. *See paper master.*

matte finish Dull paper finish without gloss or luster.

measure *In composition,* the width of type, usually expressed in picas.

mechanical A term for a camera-ready pasteup of artwork. It includes type, photos, line art, etc., all on one piece of artboard. *(page 56)*

mechanical pulp *In papermaking,* groundwood pulp produced by mechanically grinding logs or wood chips. It is used mainly for newsprint and as an ingredient of base stock for lower grade publication papers. *(page 182)*

megabyte (Mbyte, MB, Meg, or M) One million character codes on the computer. One million bytes or characters, often written MB or Mbyte. A unit of measurement equal to 1,024 kilobytes, or 1,048,576 bytes.

megahertz (MHz) Frequency equal to one million cycles per second. Measures bandwidth or analog electronic signals.

menu In *electronic publishing,* a method for selecting alternative functions displayed as a list on a workstation screen. Selection via mouse, key or sequence of keys.

metric system A decimal system adopted by most countries for solid, liquid and distance measurements. *See grammage.*

middletones The tonal range between highlights and shadows of a photograph or reproduction.

modem (MOdulator, DEModulator) A device that enables a computer to talk to other computers through phone systems by converting computer signals (data) into high-frequency voice communications signals, and vice versa.

moiré In *color process printing,* the undesirable screen pattern caused by incorrect screen angles of overprinting halftones. *(page 71)*

molleton In *offset lithography,* a thick cotton fabric similar to flannel used on the dampening rollers of a press.

monitor A video screen on a workstation.

montage In *artwork,* several photographs combined to form a composite illustration.

mottle The spotty or uneven appearance of printing, mostly in solid areas.

mouse A hand-held device that moves the cursor on a workstation by moving the device on a flat surface.

mullen tester A machine for testing the bursting strength of paper.

Mylar® In *offset preparation,* a polyester film specially suited for stripping positives because of its mechanical strength and dimensional stability.

nanometer A unit in which wavelengths of light and other radiant energy are expressed. One nanometer is one-billionth of a meter.

negative In *photography,* film containing an image in which the values of the original are reversed so that the dark areas in the subject appear light on the film and vice versa. *See positive.*

network Two or more computers which are linked and share resources to perform related tasks. Group of computers that are connected to each other by communications lines to share information and resources. *(pages 53, 109)*

newsprint Paper made mostly from groundwood pulp and small amounts of chemical pulp; used for printing newspapers.

non-impact printer An electronic device like a copier, laser or ink-jet printer that creates images on a surface without contacting it. *(page 132)*

no-screen exposure *See bump exposure.*

object-oriented An approach in drawing and layout programs that treats graphics as line and arc segments rather than individual dots. Also called vector oriented. *(page 102)*

oblong A booklet or catalog bound on the shorter dimension.

OCR (Optical Character Recognition) An electronic means of scanning (reading) copy, and converting the scanned image to an electronic equivalent. The ability to "read" printed text (characters) and convert it to digitized files that can be saved on disk and edited as a text file. *(page 91)*

off loading Relieving the intensive amount of data processing associated with a specific application (i.e., graphics) from the CPU, by performing those calculations in a dedicated or specialized processor.

off-press proofs Proofs made by photomechanical or digital means in less time and at lower cost than press proofs.

offset *See set-off. In printing,* the process of using an intermediate blanket cylinder to transfer an image from the image carrier to the substrate. Short for offset lithography.

offset gravure Printing gravure by the offset principle. Generally done on a flexographic press by converting the anilox roller to a gravure image cylinder and covering the plate cylinder with a solid rubber plate. *(page 143)*

oleophilic Oil receptive.

oleophobic Oil repellent.

opacity That property of paper which minimizes the *show-through* of printing from the back side or the next sheet.

opaque *In photoengraving and offset lithography,* to paint out areas on a negative not wanted on the plate. *In paper,* the property which makes it less transparent.

opaque ink An ink that conceals all color beneath it.

OPI (Open Prepress Interface) An extension to PostScript that automatically replaces low-resolution placeholder images with high-resolution images. *(page 98)*

orthochromatic Photographic surfaces insensitive to red but sensitive to ultraviolet, blue, green and yellow rays. *(page 64)*

overhang cover A cover larger in size than the pages it encloses.

overlay *In artwork,* a transparent covering over the copy where color break, instructions or corrections are marked. Also, transparent or translucent prints which, when placed one on the other, form a composite picture. *(page 58)*

overlay proof A color proof produced with four dyed or pigmented overlay films.

overprinting Double printing; printing over an area that already has been printed.

overrun *In printing,* copies printed in excess of the specified quantity.

packing *In printing presses,* material, usually special paper, used to underlay the image or impression cylinder in letterpress, or the plate or blanket in lithography, to get proper squeeze or pressure for printing.

page buffering The ability to spool an entire image to disk and print in a continuous motion.

page description language (PDL) *In digital prepress,* a computer language designed for describing how type and graphic elements should be produced by output devices. *(page 103)*

page makeup *In stripping,* assembly of all elements to make up a page. *(page 74) In digital imaging,* the electronic assembly of page elements to compose a complete page with all elements in place on a video display terminal and on film or plate.

pagination *In computerized typesetting,* the process of performing page makeup automatically.

palette The collection of colors or shades available to a graphic system or program.

panchromatic Photographic film sensitive to all visible colors. *(page 64)*

Pantone matching systems (PMS) Color charts that have over 700 preprinted color patches of blended inks, used to identify, display or define special colors.

paper master A paper printing plate used on an offset duplicator. The image is made by hand drawing, typewriter or electrophotography.

paste drier *In inkmaking,* a type of drier, usually a combination of drying compounds.

pasteup *See mechanical.*

PC Acronym for personal computer.

perfecting press A printing press that prints both sides of the paper in one pass through the press. *(page 133)*

PDF (Portable Document File) A proprietary format for the transfer of designs across multiple computer platforms. PDF is a universal electronic file format, modeled after the PostScript language and is device- and resolution-independent. Documents in the PDF format can be viewed, navigated, and printed from any computer regardless of the fonts or software programs used to create the original. *(page 105)*

PDL *See page description language.*

pH A number used for expressing the acidity or alkalinity of solutions. A value of 7 is neutral in a scale ranging from 0 to 14. Solutions with values below 7 are acid, above 7 are alkaline. *(page 137)*

photoconductor *In digital imaging,* materials used in electrophotography which are light sensitive when charged by corona. *(page 30)*

photomechanical *In platemaking,* pertaining to any platemaking process using photographic negatives or positives exposed onto plates or cylinders covered with photosensitive coatings. *(page 115)*

photomultiplier tube (PMT) Used in drum scanners to produce color separations of high quality. *See PMT.*

photopolymer coating *In photomechanics,* a plate coating consisting of compounds which polymerize on exposure to produce tough abrasion-resistant plates capable of long runs especially when baked in an oven after processing. *(page 120)*

pica Printer's unit of measurement used principally in typesetting. One pica equals approximately ⅙ of an inch. *(page 42)*

picking *In printing,* the lifting of the paper surface during printing. It occurs when pulling force (tack) of ink is greater than surface strength of paper.

PICT *In digital imaging,* a standard data format with which most Macintosh illustrations are encoded.

pigment *In printing inks,* the fine solid particles used to give inks color, transparency or opacity. *(page 156)*

piling *In printing,* the building up or caking of ink on rollers, plate or blanket; will not transfer readily. Also, the accumulation of paper dust or coating on the blanket of offset press.

pin register *In copy preparation,* the use of accurately positioned holes and special pins on copy, film, plates and presses to insure proper register or fit of colors. *(page 74)*

pixel Short for "picture element." A pixel is the smallest resolvable point of a raster image. It is the basic unit of digital imaging. *(page 47)*

plate cylinder The cylinder of a press on which the plate is mounted.

platesetter An image recorder which images directly on plate material. Platesetters currently available use lasers to expose or image paper, polyester or aluminum plates. *(pgs. 111 and 118)*

PMT (Photomultiplier Tube) A light-sensitive sensor that can sense very low light levels by amplifying the signals applied to it during the sensing. PMTs give drum scanners their superior color separation capabilities. *(page 92)*

point Printer's unit of measurement, used principally for designating type sizes. There are 12 points to a pica; approximately 72 points to an inch. *(page 42)*

poor trapping *In printing,* the condition in wet printing in letterpress and lithography when less ink transfers to previously printed ink than to unprinted paper. Also called *undertrapping.*

porosity The property of paper that allows the permeation of air, an important factor in ink penetration.

portrait *In photography,* vertical orientation of a format as opposed to *landscape* horizontal orientation.

position proof Color proof for checking position, layout and/or color breakout of image elements.

positive *In photography,* film containing an image in which the dark and light values are the same as the original. The reverse of negative. *(page 65)*

PostScript® A page description language developed by Adobe Systems, Inc. to describe an image for printing. It handles both text and graphics. A PostScript file is a purely text-based description of a page. *(page 103)*

preflighting *In digital prepress,* the test used to evaluate or analyze every component needed to produce a printing job. Preflight confirms the type of disk being submitted, the color gamut, color breaks, and any art required (illustrations, transparencies, reflective photos, etc.) plus layout files, screen fonts, printer fonts, EPS or TIFF files, laser proofs, page sizes, print driver, crop marks, etc. *(page 112)*

pre-press proofs *See off-press proofs.*

presensitized plate *In photomechanics,* a metal, film or paper base plate that has been precoated with a light-sensitive coating.

press proofs *In color reproduction,* a proof of a color subject made on a printing press, in advance of the production run.

pressure-sensitive paper Material with an adhesive coating, protected by a backing sheet until used.

primary colors *See additive primaries, subtractive primaries.* *(pages 61, 62)*

print quality A term describing the visual impression of a printed piece. *In paper,* the properties of the paper that affect its appearance and the quality of reproduction.

process colors *In printing,* the subtractive primaries: yellow, magenta and cyan, plus black in four-color process printing. *(page 62)*

process lens A highly corrected photographic lens with a flat field for graphic arts line, halftone and color photography.

process printing The printing from a series of two or more halftone plates to produce intermediate colors and shades.

program *In computers,* sequence of instructions for a computer. Same as *software.*

psychrometer A wet-and-dry bulb type of hygrometer. Considered the most accurate of the instruments practical for industrial plant use for determining relative humidity.

quality control is a program of activities including customer service, process control and sampling with the objective of eliminating causes of process variability now called Statistical Process Control. *(page 167)*

ragged left *In typesetting,* type that is justified on the right margin and ragged on the left.

ragged right *In typesetting,* type that is justified on the left margin and ragged on the right.

raster image processor (RIP) *In digital imaging,* a combination of computer software and hardware that controls the printing process by calculating the bitmaps of images and instructing a printing device to create the images. Most PostScript systems use a hardware RIP built into the printer. *(page 107)*

ream Five hundred sheets of paper.

red, green and blue *See RGB.*

reducers *In printing inks,* varnishes, solvents, oily or greasy compounds used to reduce the consistency for printing. *In photography,* chemicals used to reduce the density of negative or positive images or the size of halftone dots (dot etching).

reflection copy *In photography,* illustrative copy that is viewed and must be photographed by light reflected from its surface. Examples are photographs, drawings, etc. *(page 59)*

register *In printing,* fitting of two or more printing images in exact alignment with each other. *(page 58)*

register marks Crosses or other targets applied to original copy prior to photography. Used for positioning films in register, or for register of two or more colors in process printing.

relative humidity (RH) The amount of water vapor present in the atmosphere expressed as a percentage of the maximum that could be present at the same temperature.

repeatability The ability to keep photo film and the images thereon in proper register in imagesetters and film plotters. Repeatability is usually measured in micrometers.

reprography Copying and duplicating. *(page 29)*

resist *In photomechanics,* a light-hardened stencil to prevent etching of non-printing areas on plates.

resolution *In electronic imaging,* the quantification of printout quality using the number of spots per inch.

respi screen *In halftone photography,* a contact screen with 110-line screen ruling in the highlights and 220-line in the middle tones and shadows to produce a longer scale and smoother gradation of tones in the light areas of the copy. *(page 66)*

retrofit Backwards integration of advanced capability into a device or program not originally intended for that purpose.

reverse angle doctor blade *In flexography,* similar to doctor blade in gravure except used with much lighter pressure and a reverse angle on the anilox roll. *(page 141)*

RGB (Red, Green and Blue) The primary additive colors used in display devices and scanners. Commonly used to refer to the color space, mixing system or monitor in color computer graphics. *(page 61)*

right-angle fold *In binding,* a term used for two or more folds that are at 90° angles to each other. *(page 173)*

RIP *See raster image processor.*

roller stripping *In lithography,* a term denoting that the ink does not adhere to the metal ink rollers on a press. *(page 137)*

rub-proof *In printing,* an ink that has reached maximum dryness and does not mar with normal abrasion.

run-around *In composition,* the term describing type set to fit around a picture or other element of the design.

runnability Paper properties that affect the ability of the paper to run on the press. *(page 191)*

running head A headline or title repeated at the top of each page.

saddle stitch *In binding,* to fasten a booklet by wiring it through the middle fold of the sheets. Also called *saddle wire. (page 175)*

safelight *In photography,* the special darkroom lamp used for illumination without fogging sensitized materials.

scaling Determining the proper size of an image to be reduced or enlarged to fit an area. *(page 59)*

Scan-a-web *In web printing,* a rotating mirror arrangement where speed can be varied to match the speed of a press so the image on paper can be examined during printing.

scanner An electronic device used in the making of color and tone-corrected separations of images. *(pages 70* and *91)*

score To impress or indent a mark in the paper to make folding easier. *(page 172)*

screen *See contact screen.*

screen angles *In color reproduction,* angles at which the half-tone screens are placed in relation to one another, to avoid undesirable moiré patterns. A set of angles often used is: black 45°, magenta 75°, yellow 90°, cyan 105°. *(page 71)*

screened print *In photography,* a print with a halftone screen made from a halftone negative or by diffusion transfer.

screen ruling The number of lines or dots per inch on a half-tone screen. *(page 66)*

SCSI (Small Computer Systems Interface) Pronounced skuzzy, SCSI is an interface used to transmit digital data and to connect computers to peripherals. An industry-standard interface for hard drives and other storage devices that allows for very fast transfers of information.

scum *In offset lithography,* a film of ink printing in the non-image areas of a plate where it should not print.

self cover A cover of the same paper as inside text pages.

semi-chemical pulp A combination of chemical and mechanical pulping with properties similar to chemical pulp. *(page 183)*

sensitivity guide A continuous-tone gray scale with numbered steps used to control exposures in platemaking and lithfilm photography. *(page 168)*

serif The short cross-lines at the ends of the main strokes of many letters in some typefaces.

server A file server provides file data interchange between compatible peripheral devices on a local area network. Servers are identified by the type of resource they provide (e.g., disk server, file server, printer server, communications server). *(page 107)*

set-off *In presswork,* when the ink of a printed sheet rubs off or marks the next sheet as it is being delivered. Also called *offset.*

SGML (Standard Generalized Mark-Up Language) SGML is one of the newer languages for marking text for a variety of purposes, including typesetting and disk publishing. A well-designed SGML scheme enables the publisher to mark text just once for multiple uses. *(page 50)*

shadow The darkest parts in a photograph, represented in a halftone by the largest dots.

sharpen To decrease in color strength, as when halftone dots become smaller; opposite of *dot spread* or *dot gain.*

sheetwise To print one side of a sheet of paper with one plate, then turn the sheet over and print the other side with another plate using same gripper and opposite side guide.

shingling *In image assembly and layouts,* for large signatures the center or gutter margin is varied according to the position of the page in the signature and the bulk of the paper. *(page 75)*

short ink An ink that is buttery and does not flow freely.

show-through *In printing,* the undesirable condition in which the printing on the reverse side of a sheet can be seen through the sheet under normal lighting conditions.

side guide *On sheetfed presses,* a guide on the feed board to position the sheet sideways as it feeds into the front guides before entering the impression cylinder. *(page 75)*

signature *In printing and binding,* the name given to a printed sheet after it has been folded.

silhouette halftone A halftone of a subject with all of the background removed.

sizing The treatment of paper which gives it resistance to the penetration of liquids (particularly water) or vapors. *(page 183)*

skid A platform support for a pile of cut sheets of paper.

slitting Cutting printed sheets or webs into two or more sections by means of cutting wheels on a press or folder.

small caps An alphabet of SMALL CAPITAL LETTERS available in most roman typefaces approximately the size of the lower-case letters. Used in combination with larger capital letters.

soft dot Halftone dot with considerable fringe which causes dot gain or sharpening in printing or photography.

soft ink Descriptive of the consistency of paste inks. *(pg. 156)*

soft proof *See hard proof.*

software *See program.*

spectrophotometer Instrument for measuring color for CIE color spaces. It is more accurate than most color colorimeters.

spectrum The complete range of colors in the rainbow, from short wavelengths (blue) to long wavelengths (red). *(page 84)*

SPC Acronym for Statistical Process Control. *(page 167)*

spine *See backbone.*

spiral binding A book bound with wires in spiral form inserted through holes punched along the binding side. *(page 178)*

spool (simultaneous peripheral operations on-line) Refers to an output data set that is waiting for a print device. *(page 108)*

spot Smallest visible point that can be displayed or printed. The smallest diameter of light that a scanner can detect, or an image-setter or printer can image. Dot should not be confused with spot. *(page 47)*

spreads and chokes *See chokes and spreads.*

star target Film pinwheel used to measure resolution of plates during production and degradation during printing. *(page 168)*

static neutralizer *In printing presses,* an attachment designed to remove the static electricity from the paper to avoid ink set-off and trouble with feeding the paper.

step-and-repeat *In photomechanics,* the procedure of multiple exposure using the same image by stepping it in position according to a predetermined layout or program. *(page 117)*

stet A proofreader's mark, written in the margin, signifying that copy marked for corrections should remain as it was.

stochastic screening A digital screening process that converts images into very small dots (14-40 microns) of equal size and variable spacing. Second order screened images have variable size dots and variable spacing. Also called Frequency Modulated (FM) screening. *(page 72)*

stock Paper or other material to be printed.

stone *In lithography,* formerly used as the plate material and presently used by artists as an art medium. *In letterpress,* the bed on which metal type is leveled and locked up.

strike-on composition or cold type Type set on typewriter composing machines. *(page 46)*

strike-through *See show-through.*

stripping *In image assembly,* the positioning of negatives (or positives) on a flat to compose a page or layout for platemaking. *(page 74) In printing,* ink stripping on ink rollers prevented by plastic or copper coated steel rollers in the ink roller train. *(page 137)*

substance The weight in pounds of a ream (500 sheets) of paper cut to the standard size (17″ x 22″) for business papers (bond and ledger): e.g., 20 pounds. Similar to basis weight of other grades of paper. *(page 189)*

substrate Any material that can be printed on, such as paper, plastic and fabric.

subtractive primaries Yellow, magenta and cyan, the hues used for process color printing inks. *(page 62)*

sulphate pulp Paper pulp made from wood chips cooked under pressure in a solution of caustic soda (sodium hydroxide) and sodium sulphide. Known as kraft. *(page 182)*

sulphite pulp Paper pulp made from wood chips cooked under pressure in a solution of bisulphite of lime (calcium bisulphite). *(page 182)*

supercalender *In papermaking,* a calender stack, separate from the papermaking machine, with alternate metal and resilient rolls, used to produce a high finish on paper. *(page 185)*

supercell *In digital halftone imaging,* a combination of sub-groups of halftone dots that are handled as a single group.

surprint *In photomechanics,* exposure from a second negative or flat superimposed on an exposed image of a previous negative or flat.

SWOP Specifications for Web Offset Publications. *(page 169)*

tack *In printing inks,* the property of cohesion between particles; the separation force of ink needed for proper transfer and trapping on multicolor presses. A tacky ink has high separation forces and can cause surface picking or splitting of weak papers. *(page 157)*

Tackoscope *See inkometer.*

tagged image file format (TIFF) A file format for graphics suited for representing scanned images and other large bitmaps. TIFF is a neutral format designed for compatibility with all applications. TIFF was created specifically for storing gray-scale images, and it is the standard format for scanned images such as photographs — now called TIFF/IT. *(page 104)*

terabyte (TB) One trillion bytes.

text The body matter of a page or book, as distinguished from the headings.

the internet A network of networks that links workstations over telecommunication lines to share files and exchange e-mail internationally. *(page 53)*

thermal dye sublimation Like thermal printers, except pigments are vaporized and float to desired proofing stock. Similar to Thermal Dye Diffusion Transfer, or D2T2. *(page 153)*

thermal transfer printers These printers use a *transfer* sheet that carries ink in contact with the paper or transparency, and a heated printhead driven by digital data that touches the transfer sheet to transfer images to the right points on the page. *(page 153)*

thermo-mechanical pulp *In papermaking,* made by steaming wood chips prior to and during refining, producing a higher yield and stronger pulp than regular groundwood. *(page 183)*

thixotropy False body in inks. *(page 156)*

TIFF *See tagged image file format.*

tints Various even tone areas (strengths) of a solid color.

tissue overlay A thin, translucent paper placed over artwork (mostly mechanicals) for protection; used to indicate color break and corrections.

tolerances The specification of acceptable variations in register, density, dot size, plate or paper thickness, concentration of chemicals and other printing parameters. *(page 169)*

toner *In digital printing,* imaging material also called digital inks, used in plateless printing systems like electrophotography, magnetography, ion or electron deposition and laser printers *(page 165). In inks,* dye used to tone printing inks, especially black.

tone reproduction The tonal relationship between all the elements of a reproduction. *(page 67)*

toning *See scum.*

tooth A characteristic of paper, a slightly rough finish, which permits it to take ink readily.

transparency Color positive film.

transparent copy *In photography,* illustrative copy such as a color transparency or positive film through which light must pass in order for it to be seen or reproduced.

transparent ink A printing ink which does not conceal the color beneath. Process inks are transparent so that they will blend to form other colors.

transpose To exchange the position of a letter, word or line with another letter, word or line.

trapping *In printing,* the ability to print a wet ink film over previously printed ink. *Dry trapping* is printing wet ink over dry ink. *Wet trapping* is printing wet ink over previously printed wet ink. *In prepress,* refers to how much overprinting colors overlap to eliminate white lines between colors in printing. *See spreads and chokes. (page 112)*

trim marks *In printing,* marks placed on the copy to indicate the edge of the page.

twin-wire machine *In papermaking,* a fourdrinier paper machine with two wires instead of one producing paper with less two-sidedness. *(page 184)*

two-sheet detector *In printing presses,* a device for stopping or tripping the press when more than one sheet attempts to feed into the grippers.

two-sidedness *In paper,* the property denoting difference in appearance and printability between its top (felt) and bottom (wire) sides. *(page 187)*

type gauge *In composition,* a printer's tool calibrated in picas and points used for type measurement.

UCA (UnderColor Addition) *In process color printing,* used with GCR, UCA is ink added in shadow areas to increase color saturation. *(page 69)*

UCR (UnderColor Removal) *In process multicolor printing,* color separation films are reduced in color in neutral areas where all three colors overprint and the black film is increased an equivalent amount in these areas. This improves trapping and can reduce makeready and ink costs. *(page 69)*

UGRA test target A measure of image resolution and dot size on plates and in printing. *(page 168)*

undercut *In printing presses,* the difference between the radius of the cylinder bearers and the cylinder body, to allow for plate (or blanket) and packing thickness.

unit *In multicolor presses,* refers to the combination of inking, plate and impression operations to print each color. A 4-color press has 4 printing units each with its own inking, plate and impression functions.

UNIX A multiuser, multi-tasking operating system that runs on a wide variety of computer systems from micro to mainframe. UNIX was written in the C programming language. It is the most common operating system for servers on the Internet. *(page 97)*

-up *In printing,* two-up, three-up, etc., refers to imposition of material to be printed on a larger size sheet to take advantage of full press capacity.

upper case Capital letters in type. *(page 36)*

UV inks *In printing,* solventless inks that are cured by UV radiation. They are used extensively in screen printing, narrow web letterpress and flexographic printing. *(page 161)*

vacuum frame *In platemaking,* a vacuum device for holding copy and reproduction material in contact during exposure.

varnish A thin, protective coating applied to a printed sheet for protection or appearance. *(page 164) Also, in inkmaking,* it can be all or part of the ink vehicle.

VDT *See video display terminal.*

vehicle *In printing inks,* the fluid component which acts as a carrier for the pigment.

vellum finish *In papermaking,* a toothy finish which is relatively absorbent for fast ink penetration. *(page 187)*

video display terminal (VDT) A term for a computer monitor or display. *(page 91)*

vignette An illustration in which the background fades gradually away until it blends into the unprinted paper. *(page 81)*

viscosity *In printing inks,* a broad term encompassing the properties of tack and flow. *(page 156)*

WAN (Wide Area Network) Any internet or network that covers an area larger than a single building or campus. A collection of disparate, widely located and geographically isolated networks, connected by private or public communication lines. *(page 110)*

warm color *In printing,* a color with a yellowish or reddish cast.

washup The process of cleaning the rollers, form or plate, and sometimes the ink fountain of a printing press.

waterless plate *In platemaking,* printing plate with silicone rubber coating in non-image areas, that is printed on an offset press without dampening solution. *(page 121)*

waterless printing *In offset,* printing on a press using special waterless plates and no dampening system. *(page 138)*

web A roll of paper used in web or rotary printing.

web press A press which prints on a roll of paper.

web tension The amount of pull or tension applied in the direction of travel of a web of paper by the action of a web press.

widow *In composition,* a single word or part of a word on a line by itself, ending a paragraph, or starting a page, frowned upon in good typography.

wire-o binding A continuous double series of wire loops run through punched slots along the binding side of a booklet.

wire side *In papermaking,* the side of a sheet next to the wire in manufacturing; opposite from felt or top side. *(page 187)*

with the grain Folding or feeding paper into a press with the grain of the paper parallel to the blade of the folder or the axis of the impression cylinder.

woodcut An illustration in lines of varying thickness, cut in relief on plank-grain wood, for the purpose of making prints by a relief printing method like letterpress. *(page 14)*

word processor A typewriter connected to a computerized recording medium to input, edit and output digital text data. *(page 51)*

work-and-tumble To print one side of a sheet of paper, then turn it over from gripper to back using the same side guide and plate to print the second side. *(page 75)*

work-and-turn To print one side of a sheet of paper, then turn it over from left to right and print the second side using the same gripper and plate but opposite side guide. *(page 75)*

WORM (Write Once Read Many Times) A type of optical memory device.

wove paper Paper having a uniform unlined surface and a soft smooth finish.

wraparound plate *In rotary letterpress,* a thin one-piece relief plate which is wrapped around the press cylinder like an offset plate. Can be used for direct or indirect (offset) printing. *(page 125)*

wrinkles Creases in paper occurring during printing. *In inks,* the uneven surface formed during drying.

wrong font *In proofreading,* the mark "WF" indicates a letter or figure of the wrong size or face. *(page 43)*

WWW (World Wide Web) The highly inter-connected network of hypertext servers (HTTP servers) which allow text, graphics, sound and video files to be displayed. *(page 53)*

WYSIWYG (What You See Is What You Get) Means that what you see on the computer monitor is generally the same as what appears on the hard copy or film. Pronounced "wizzywig". *(page 100)*

xerography An electrophotographic copying process that uses a corona charged photoconductor surface, electrostatic forces and dry or liquid toner to form an image. *(page 30)*

yellow Hue of a subtractive primary and a four-color process ink. It reflects red and green light and absorbs blue light. *(page 62)*

International Paper Printing Papers

HAMMERMILL® MAKES YOUR WORK LOOK BETTER™

Springhill®

Beckett

Strathmore

INTERNATIONAL PAPER

International Paper, founded in 1898, is the world's largest forest products company and employs more than 117,000 people worldwide. It is a worldwide producer of printing and writing papers, paperboard and packaging products and wood products. The company also operates specialty products businesses and distributes paper and wood products. International Paper has manufacturing operations in approximately 50 countries and exports its products to more than 130 nations.

International Paper's fine printing and office papers are marketed under the Beckett, Carolina, HammerMill, Springhill, Strathmore and Zanders brand names and can be found through an extensive nationwide network of leading paper merchants. International Paper also distributes its paper through xpedx, a division of the company that includes more than 250 retail stores.

International Paper is one of the leading U.S. producers of recycled-content printing and writing papers with more than 100 grades on the market, many of which are listed on the following pages.

FINE PAPERS

BECKETT PAPERS

Beckett Papers is an industry pioneer. Founded in 1848, Beckett is the oldest operating fine paper mill west of the Allegheny Mountains and the fourth oldest operating mill in the United States. Beckett began its operation by making 100 percent rag content newsprint. In 1894, after experimenting with new aniline dyes, Beckett developed the industry's first extensive range of colored cover papers.

In the years that would follow, Beckett Papers can take credit for an unequaled list of industry innovations. The first premium lithographic paper, the first fluorescent lithographic paper, the first use of corrugated cartons, the first use of polyethylene-lined cartons and the first use of polyethylene wrap on skids are among such innovations.

After HammerMill Papers acquired Beckett in 1959, the company introduced lightweight duplex papers to the industry. It took advantage of its expertise in embossing by being the first to introduce a comprehensive line of linen printing and writing papers — Cambric.

The commitment of Beckett Papers to be the first fine paper mill to manufacture its complete product line with recycled fiber is the latest in this long list of innovations.

Beckett® Cambric Writing, Text and **Cover** is one of the most prestigious recycled linen grades on the market. Since its introduction in 1970 as the first comprehensive linen grade, Cambric has constantly maintained a leadership position. Cambric, which means linen, has brand awareness and an extraordinary reputation for high quality and on-press performance. Its subtle linen pattern and extensive selection of colors and basis weights offer more design options than any other premium recycled linen grade.

Cambric is 20 percent recycled, including 20 percent post-consumer waste based on fiber weight. This watermarked grade has a soft, subtle, two-side embossed linen finish that enhances printed images without overpowering and provides excellent ink holdout for superior printability and even-sided tactile appeal. Cambric has two bright, clean whites and a full range of colors. Cambric is guaranteed for photocopiers, laser and inkjet printers and has exceptional strength, making it suitable for scoring, folding and three-dimensional techniques. It is available in 24 lb. Writing, 70 and 80 lb. Text, 80 and 100 lb. Cover, Duplex and Double Thick Cover.

Beckett Concept® Writing, Text and **Cover** was introduced in 1990 as one of the first recycled writing, text and cover grades. Concept has a natural, earthy color palette that prints beautifully and provides unlimited design options. Beckett Concept provides outstanding versatility while addressing the need for making better use of the natural resources around us.

Concept is 30 percent recycled, including a minimum of 30 percent post-consumer waste based on fiber weight. This acid-free, watermarked paper includes two new bright whites: Radiance, our non-recycled 98 bright white, Polar, a 30 percent recycled 95 bright white and four new bold accent colors. Direct-to-plate sizes of Concept are available. Concept Writing and Text are guaranteed for use in photocopiers, laser and inkjet printers. Concept is available in 24 lb. Writing, 70 lb. Text and 80, 100 and 130 lb. Cover and is known for exceptional on-press performance.

Beckett Enhance® Writing, Text and **Cover** is the premier silk finish recycled grade on the market and a leader in innovation since its introduction. With its marbleized items and trend-setting color palette, Enhance echoes a unique excitement that can make any printed piece jump to life.

Enhance is 20 percent recycled, including 20 percent post-consumer waste based on fiber weight. This watermarked grade has a soft, subtle, two-side embossed satin finish that enhances printed images without overpowering and provides excellent ink holdout for superior printability and even-sided tactile appeal. Enhance has a bright, clean white and a full range of fashionable colors which include a variety of marbleized items. Writing and text weights are guaranteed for use in photocopiers, and laser and inkjet printers and Enhance has exceptional strength, making it suitable for scoring, folding and three-dimensional techniques. Enhance is available in 24 lb. Writing, 70 and 80 lb. Text, 80 and 100 lb. Cover.

Beckett Expression® Writing, Text and **Cover** was introduced in 1993 and has been tremendously successful due to its exceptionally smooth surface, sophisticated color palette, recycled content and economical price.

Expression is 30 percent recycled, including a minimum of 30 percent post-consumer waste based on fiber weight. This acid-free grade is manufactured to have an exceptionally smooth finish and is available in a sophisticated selection of soft, versatile colors, including Radiance, a new 98 bright white non-recycled sheet. Expression is designed for excellent on-press performance to produce fine detail with minimal dot gain, whether it be halftones, duotones or four-color process. Expression Writing and Text are guaranteed for use in photocopiers, laser and inkjet printers. In 1999, digital (xeikon qualified) rolls and direct-to-plate sizes were added to Expression. Expression is available in 24 lb. Writing, 70, 80 and 100 lb. Text and 65, 80, 100 and 130 lb. Cover.

Beckett Paradox® Text and **Cover** was introduced in 1997. It is available in two finishes: a felt finish and a matching smooth finish. Paradox has a bright, clean, recycled white and a range of contemporary colors. Paradox is 20 percent recycled, including a minimum of 20 percent post-consumer waste based on fiber weight. Paradox colors are all manufactured acid free. It is ideal for annual reports, brochures, corporate identity packages, press kits, greeting cards, invitations and manuals. Paradox is available in 70 and 80 lb. Text and 65, 80 and 100 lb. Cover.

Beckett Ridge® Text, Cover and **Duplex** has a unique linear finish developed by Beckett in 1982. This grade combines an exceptional blend of individuality and flexibility. The distinct texture of Beckett Ridge can be used in a variety of printing and design applications that stimulate the imagination. Its embossed finish will not interfere with delicate type or solid coverage, ensuring a beautifully designed finished product that is both unique and appealing. Ridge is 20 percent recycled, including 20 percent post-consumer waste based on fiber weight. The distinct and unique linear embossed finish enhances printed images and provides excellent ink holdout for superior printability. Ridge is a versatile grade with exceptional strength, making it suitable for scoring, folding and three-dimensional techniques. It has two bright, clean whites and a full range of colors. It is available in 80 lb. Text, 80 and 100 lb. Cover and Duplex.

Beckett R.S.V.P.™ **Text, Cover** and **Duplex** is a recycled, true felt grade that lends itself exceptionally well to finishing techniques such as embossing, foil-stamping, die-cutting and thermography. It combines a rich, traditional look with a prestigious finish.

R.S.V.P. is 20 percent recycled, including a minimum of 20 percent post-consumer waste based on fiber weight. It has a rich true felt finish and is available in 10 beautiful colors. It is a versatile grade with exceptional strength, making it well suited for scoring, folding and three-dimensional techniques. It is crafted to have a high bulk characteristic, which gives this sheet a substantial look and rich tactility. It is even-sided, which guarantees top performance on both felt and wire sides and makes it ideal for work and turn printing. It is available in 80 lb. Text, 80 lb. Cover, 100 lb. Cover and Duplex in a variety of sizes.

STRATHMORE PAPERS

In the 100+ years since its inception, Strathmore quality has become known throughout the world. Strathmore manufactures and markets papers in three distinct categories: Bond & Writing, Text & Cover and Artist Products. Today, Strathmore is the U.S. leader in 25 percent cotton fiber Writing paper and is recognized for its distinctive Text & Cover papers in a variety of colors and finishes. In addition, Strathmore Artist Products have been a first choice for both professional and amateur artists who value quality.

THE STRATHMORE WRITING SYSTEM

Strathmore manufactures premium cotton fiber Bond & Writing papers with rich surfaces, distinctive whites and soft colors, and a range of cotton fiber contents, which add elegance to business and personal communications. In addition, most Strathmore Bond & Writing grades include post-consumer fiber content for papers that are as environmentally friendly as they are beautiful.

The most complete system of business correspondence papers available, the Strathmore Writing System allows businesses to match paper throughout their entire printed communications system including letterhead, labels, business cards, executive stationery and even collateral material. In addition, of course, matching envelopes are available in each Strathmore Writing color and type.

Strathmore Writing® is the 25 percent cotton correspondence paper most often specified in America. The cotton fiber adds strength, enhances the feel and look of the paper and conveys the image of quality. Available in four core whites and five sophisticated colored neutrals. It is available recycled, with 30 percent post-consumer waste, and in shades of white, as it also is available with virgin fiber. All recycled and virgin grades match each other visually and demonstrate the craftsmanship and quality for which Strathmore is known. Available in Wove and Laid finishes, it features the Strathmore laser and inkjet guarantee. Acid free.

Strathmore® **Pure Cotton** is the highest quality correspondence paper containing 100 percent cotton fiber with a distinctive localized watermark. This all-cotton grade offers a luxurious feel, authoritative crispness and unparalleled strength and permanence that demonstrates the ultimate in craftsmanship and quality. Strathmore Pure Cotton is available in four whites that coordinate with the Strathmore Writing System. Available in Wove and a special Laid finish, it features the Strathmore laser and inkjet guarantee. Acid free.

Strathmore Script™ is a high-performance recycled correspondence paper containing 30 percent post-consumer waste. It is available in Smooth, Very Smooth, Pinstripe, Laid and Linen finishes in four whites and two colors that coordinate with the Strathmore Writing System. In addition, Script now has an expanded palette of nine contemporary colors. This sheet demonstrates the craftsmanship and quality for which Strathmore is known at a very accessible price point. Features the Strathmore laser and inkjet guarantee. Acid free.

Strathmore Writing® **Cover Bristol** is a premium recycled pasted sheet with virgin fiber options available in the whites. Pasting two base sheets provides extra snap and rigidity, dimensional stability and the same finish on both sides. It is available in nine colors including four whites. Cover 88 Pasted comes in Wove and Laid finishes, in all the Strathmore Writing whites and colors, and in a Plate finish in the four whites. Cover 110 Paste is available in whites and colors in the Wove finish and is acid free. The recycled offerings contain 30 percent post-consumer waste. This paper is ideal for business cards, folders and any other applications requiring a strong, rigid surface that performs well for all finishing processes.

Strathmore Writing® Cover is a premium recycled cover stock containing 30 percent post-consumer waste—except for Ultimate White, which is made with virgin fiber for the brightest possible white. This paper is ideal for business cards, folders and any other applications requiring a strong, rigid surface that performs well for all finishing processes. Available in Wove and Laid finishes. For a matching smooth finish, refer to Strathmore Elements. Acid free.

Strathmore Writing® Text is a premium text sheet available in all the Strathmore Writing colors. It contains 30 percent post-consumer waste, except for Ultimate White, which is made with virgin fiber for the brightest possible white. Strathmore Writing Text demonstrates the craftsmanship and quality for which Strathmore is known and it provides an excellent surface for a wide range of printing and production processes. For a matching smooth finish, refer to Strathmore Elements. Available in a rich, velvet Wove and a visual, tactile Laid finish. Acid free.

Strathmore Writing® Label Stock is a pressure-sensitive, permanent adhesive stock for labels, stickers and tip-ons of all kinds. It contains 30 percent post-consumer waste, except for Ultimate White, which is made with virgin fiber for the brightest possible white. Providing superior printability, opacity and excellent stiffness for kiss-cut applications, Strathmore Writing Label Stock is available in four whites in a Wove finish. The label stock will match, in color and finish, the other components of the Strathmore Writing System. Suitable for most desktop inkjet and laser printers. Acid free.

Strathmore Script™ Text and **Cover** papers are high performance recycled papers with 30 percent post-consumer waste, except for Ultimate White. In addition to the four core whites of the Strathmore Writing System, Script Text is available in 14 contemporary colors. Available in Smooth, Very Smooth, Pinstripe, Laid and Linen Finishes. The 21 colors of Script are organized by color family—Whites, Creams, Warm Tones, Cool Tones, Neutrals, Brights and Deep Tones. The expansion of Script offers the most complete correspondence line of paper for today's style-conscious market places. Acid free.

BOND AND WRITING GRADES

Strathmore Bond® has set the standard of excellence in 25 percent cotton fiber paper since the 1930s. Available in two distinctive whites.

Strathmore Renewal® Writing is the next generation of recycled papers with 30 percent post-consumer waste, perfect for multiple printing mediums. Renewal's palette ranges from whites and naturals to pleasing midtones and a collection of bolds that are unique among recycled papers. The fibered selections offer subtle visual textures that add sophistication and aesthetic appeal to any printed piece. Renewal's surface has been reengineered for greater smoothness. It is laser and inkjet guaranteed. Acid free.

TEXT AND COVER GRADES

Strathmore is known for the rich textures and wide range of unique colors of its Text and Cover papers. Strathmore takes the time and care to craft its Text and Cover papers with authentic marking felts that gently mold the fibers into surfaces that are visually and tactilely appealing. In addition, a growing number of Strathmore Text and Cover papers contain pre- or post-consumer recovered fiber.

Strathmore Americana® Text and **Cover** is a premium authentic felt-finished paper that is truly original in both texture and color. An elegant, distinctive texture is achieved by combining a dandy roll watermark with a genuine marking felt. Its unique surface is especially suitable for embossing while receptive to all printing and production processes, from four-color process, engraving and foil stamping to diecutting.

Strathmore Beau Brilliant® Text and **Cover** is a premium felt-finished paper that offers the deepest, most widely recognized texture in the industry. Its distinctive, deep texture is achieved with a genuine marking felt. The unique surface is suitable for all printing and production processes from four-color process, engraving and foil stamping to die cutting, embossing and scoring. The deep texture of Beau Brilliant makes it ideal for all types of high impact cover applications.

Strathmore Elements® is a premium recycled paper available in new solid colors in addition to six innovative, patterned surface treatments. The exceptionally smoother, level surface offers excellent printability. Contains 20 percent post-consumer waste, except for Ultimate White. Patterns are created with papermaking dyes applied by a unique manufacturing process. It is an ideal choice for all printing and finishing processes, from four-color process, engraving and foil stamping to diecutting, embossing and scoring. Text and writing weight items carry the Strathmore laser guarantee. Includes a new, 110 lb. Cover for extra bulk and rigidity in high-impact cover uses.

Strathmore Fiesta® Text and **Cover** is a premium authentic felt-finish paper that offers a unique, colored deckle edge. The soft pastel shades of the paper provide a contrast to the rich colors of the deckle edge. The elegant texture is achieved with genuine marking felts at the wet end of the papermaking process, which contrasts with the feather deckle edge. The unique colored deckle provides a two-color effect before printing. Excellent for announcements and cards where the deckle on the envelope flap adds impact.

Strathmore Grandee® has a rich texture achieved by combining a genuine marking felt with a dandy roll mark. Exceptional surface for all printing and production processes, from four-color process, engraving and foil stamping to die cutting, embossing and scoring. Outstanding ink holdout and opacity make it appropriate for all types of distinctive communications. The colors of Grandee are rich, contemporary and tasteful, enhancing the impact of marketing communication pieces. Contains 20 percent post-consumer waste.

Strathmore Pastelle® Text and **Cover** is a premium authentic felt-finish paper, which offers an elegant deckle edge. The elegant, subtle, uniform texture is achieved by genuine marking felts at the wet end of the papermaking process. The deckle edge adds impact and elegance to printed pieces, and is especially effective on envelope flaps. Excellent surface is suitable for all printing and production techniques, from four-color process, engraving and foil stamping to diecutting, embossing and scoring. Impressive ink holdout and opacity makes it ideal for all types of social correspondence and high-end marketing communications.

Strathmore Renewal® Text and **Cover** is the next generation of recycled papers with 30 percent post-consumer waste, perfect for multiple printing mediums. Renewal's palette ranges from whites and naturals to pleasing midtones and a collection of bolds that are unique among recycled papers. The fibered selections offer subtle visual textures that add sophistication and aesthetic appeal to any printed piece.

Strathmore Rhododendron® Text and **Cover** is a premium paper with 25 percent cotton content and a Telanian finish. The elegant Telanian finish duplicates the texture of hand-woven Irish linen. Its exceptional surface is suitable for all printing and production processes, from four-color process, engraving and foil stamping to diecutting, embossing and scoring. Outstanding ink holdout and opacity make it ideal for all types of distinctive communication. The cotton content adds strength and durability to this enduring grade in its multiple cover weight options.

ZANDERS PAPERS

The rich history of Zanders papers began in 1829 when Johann Wilhelm Zanders founded the paper mill that has developed into today's Zanders Feinpapiere AG. Over the last 166 years, the German paper manufacturer has adapted to meet the changing needs of the market.

From the beginning, the name Zanders has been synonymous with quality. With two production plants in operation today, Zanders' philosophy is to produce the finest paper with the highest technical and aesthetic qualities.

Zanders papers were first introduced to the American market in 1978. By the time of Zanders' 150th anniversary in 1979, the company was exporting paper to more than 100 countries worldwide.

CAST COATED

In general, the cast coating process produces an unparalleled printing surface. Among cast coated papers, Chromolux offers additional benefits, the results of years of research and development. First, the superior whiteness of Chromolux permits faithful color reproduction. Secondly, its unique surface offers superior ink holdout so that there is little or no loss of ink gloss on printed image areas. Finally, the Chromolux coating has been specially formulated for maximum flexibility. It combines with the generous bulk and stiffness of the base stock to resist cracking when properly scored, either with or against the grain.

Chromolux® C1S is cast coated on one side and matte coated on the reverse. It is available in four cover and two label weights.

Chromolux® C2S is cast coated on two sides and is available in three cover weights.

Chromolux® Metallic Cover is a unique product created when the finest paper mill in Europe bonds a lacquer based pigment to the incomparable Chromolux base stock. It is available in 13 true metallic colors, all having a snowy white matte coated reverse.

Chromolux® Color Cover is created using the same innovative technology as Chromolux Metallic Cover. It is available in a dazzling range of 11 shades, especially created for quality jobs requiring full coverage on a cast coated cover.

Chromolux® Embossed Cover is a superior cast coated stock with the added appeal of an embossed surface. The rich colors offered eliminate the need for printing a background color while the linear embossing provides both visual and tactile appeal.

Chromolux® Vario delivers the finest cast coated surface available and simultaneously relieves the printer of the need for printing a background color on the reverse. It is highly appropriate for projects using a cast coated cover with solid color coverage on the matte side. Startling duplex results may be obtained by the creative use of die cuts and folds.

Chromolux® Mirri is a glamorous specialty cover with a unique reflective surface and a matte coated reverse side. This distinctive paper is available in a stunning array of brilliant colors and a versatile range of three weights, identified as Mirriwrap, Mirricard and Mirriboard.

ENAMEL COATED

Ikono is the premium enamel that sets the world standard for high quality coated paper. Its exceptionally smooth surface provides the perfect foundation for level ink lay and consistent solids.

Ikono® Gloss offers high whiteness, a 96 brightness and a 76 gloss for brilliant reproduction, excellent ink holdout and high printed ink gloss. Ikono Gloss contains 20 percent post-consumer waste with a 50 percent total recycled fiber content. Designers and printers can rely on this paper to meet current environmental standards without sacrificing any optical or performance characteristics.

Ikono® Dull Satin has a semi-gloss finish with a 96 brightness. Its low luster surface offers easy readability yet delivers superior ink holdout normally associated with glossier papers. Ikono Dull Satin contains 20 percent post-consumer waste with a 50 percent recycled fiber content.

Ikono® Dull Satin Cream combines the same surface characteristics of Ikono Dull Satin with a refreshing, creamy shade. A perfect choice for fine art books, museum catalogs and illustrated books, its soft color evokes a mood of tradition and elegance.

Ikono® Matt combines a high white, non-reflective surface with an inviting velvety feel. Its 96 brightness permits faithful color reproduction while enhancing the contrast between image and non-image areas.

Zanders® Mega is a proven, value-added high performance enamel-coated paper. This sheet contains 20 percent post-consumer waste with a 50 percent total recycled fiber content.

Zanders® Mega Gloss offers high whiteness and a 92 brightness as the foundation for consistent reproduction and dependable renewability.

Zanders® Mega Dull is a "true dull" providing maximum contrast between paper surface and printed ink gloss. This combined with the sheet's high whiteness and 92 brightness gives it excellent printability.

ZANDERS ON-DEMAND

Zanders On-Demand products answer the need for quality coated papers suited to the smallest digital presses and higher resolution color copiers that have become prevalent with the advent of digital printing. These papers provide quality and performance for outstanding color work on today's leading equipment.

Zanders Imaging® *digital* consists of rolls for web-fed digital presses. With a high brightness, superior whiteness and exceptionally smooth surface, these papers make color graphics look better than ever. Both Zanders Imaging *digital* Supergloss and Zanders Imaging *digital* Dull are splice-free and fully comply with all current web-fed digital press requirements. Both offer dependable performance and exciting results. These papers are ideal for top-quality, short-run color and customized communications.

Zanders Imaging® *copy* is a bright white paper that enhances the color brilliance, contrast and definition achieved on color copy machines. Offering a selection of high quality coated papers especially designed for today's full-color copying systems, these papers provide exacting reproduction rich in contrast, with excellent color gradation and sharp contours. Ideally suited for proof copies, presentation material and self-created brochures, Zanders Imaging *copy* offers perfect toner adhesion and optimal utilization of a print system's performance range.

Zanders *print-on-demand* offers cut-sizes of cast coated Chromolux C1S and enamel coated Zanders Mega in small quantities to give short runs affordable excellence. This conveniently packaged availability means that even limited quantities of materials can be produced on paper that sets the world standard for quality and performance.

Panorama® is a wide format inkjet offering for signage, engineering drawings, proofing and other short-run, large format applications.

SPECIALTY PAPERS

Zanders T-2000 RO is a highly translucent, natural tracing paper with a fine, cloudless formation. Its special surface treatment (RO) ensures clearly defined drawings by pencil or pen with no ink strike-through. When used as a protective overlay, the special surface treatment prevents migration of ink from adjacent sheets into the paper's surface. It will not yellow and will maintain its transparency over time.

Elephant Hide® Paper is an extraordinary product combining timeless beauty with uncommon practicality. The characteristic veining varies sheet to sheet, as do the shade and caliper, making the resulting product as close to handmade as possible, with a striking resemblance to marble. Yet, owing to a special surface treatment, the paper has a high degree of resistance to scratching, scuffing and grease stains, and it can be wiped with a damp cloth. Recognized worldwide for its unusual surface and rare properties, Elephant Hide Paper is used for documents, menus, labels, envelopes, greeting cards and as a book-binding material for sturdy volumes as well as small brochures.

COATED PAPERS

Preference® is a perfect No. 1 coated web freesheet. Sparkling and brilliant with an icy-smooth, flat surface and exceptional print gloss. Bright blue-white shade. Guaranteed to perform. Available in Gloss or Dull in basis weights 70 to 100 lb.

Savvy™ is a No. 2 coated web freesheet. This grade offers outstanding qualities in every important paper attribute: incomparable print and paper gloss, balanced blue-white shade and high brightness. It will change the way you look at No. 2 coated web papers. Available in Gloss or Matte basis weights 60 to 100 lb.

Accolade® Gloss is a superb No. 3 coated web offset paper that looks and prints like a No. 2. This product is a dull sheet designed for offset printing with excellent rotogravure print results. Available in 40 to 100 lb., Accolade Gloss is characterized by a high print gloss, uniform ink lay, high brightness, a clean film-free surface and outstanding press performance. Accolade Gloss is the right choice for catalogs, publications or commercial print jobs. Available also in recycled 45 to 100 lb. with 10 percent post-consumer waste.

Accolade® Matte is a No. 3 coated free web offset paper designed to provide a softer appearance with low paper gloss. Available in 45 to 100 lb., Accolade Matte features uniform ink lay, higher bulk and higher brightness. Ideal for books, magazines, catalogs and commercial print jobs where glare must be minimized and bulk is important. Accolade Matte is available in recycled with 10 percent post-consumer waste.

Accolade® Dull is a No. 3 coated free web offset paper designed with a satin finish and reduced paper gloss for ease of reading. This product is available in 45 to 100 lb. Great dot fidelity and print uniformity make this the choice for any commercial print application requiring a glare-free look. Accolade Dull is available in recycled with 10 percent post-consumer waste.

Influence® is a No. 3 coated web freesheet. This grade leads in all measurements of paper quality: opacity, surface uniformity and gloss. Exceptional versatility for on-press ease. The quality you expect from a No. 2 at a No. 3 price. Blue-white shade. Guaranteed to perform. Available in Gloss and Soft Gloss in basis weights 45 to 80 lb., Matte in basis weights 50 to 80 lb. and Gravure in basis weights 45 to 70 lb.

Velocity™ Gloss is a No. 3 coated web freesheet. This perfect house sheet is engineered for runnability and flawless press performance—roll after roll, run after run. Balanced shade. Available in basis weights 50 to 100 lb.

Liberty 2000™ is a No. 4 coated groundwood web offset paper featuring a bright, blue white look of unparalleled, uniform whiteness. Available in 38 to 60 lb., Liberty 2000 is designed to provide unique excellence in ink lay smoothness, dot fidelity and halftone cleanliness to yield high print contrast. Available also in recycled with 10 percent post-consumer waste.

Liberty 2000 G™ is a No. 4 coated groundwood web gravure grade featuring superb whiteness and dot reproduction characteristics. This product is available in 38 to 50 lb. basis weights. Liberty 2000 G has excellent rotogravure print gloss with minimum dot deletion. Liberty 2000 G's excellent print performance lends itself to all your catalog and publication needs.

Hudson® Web Gloss is a No. 5 coated groundwood web offset grade. Available in 30 to 60 lb. basis weights. Designed for great print uniformity, paper surface appearance and consistency, this product is a steady performer for catalogs, magazines, freestanding inserts, coupons and other commercial print jobs requiring lightweight to medium weight groundwood. Available also in recycled with 10 percent post-consumer waste.

Maineweb® is a No. 5 coated groundwood web offset paper. This grade is synonymous with reliability, runnability and versatility. With excellent print quality, it is the benchmark of the industry. For heatset web offset presses. Available in 32 to 34 lb.

Mainebrite® is a No. 5 coated groundwood web offset paper. This grade is known for high-brightness, high gloss and dependable runnability for demanding high-quality print jobs. For heatset web offset presses. Available in basis weights 36 to 40 lb.

Mainebulk® is a No. 5 coated groundwood web offset paper. This grade offers unique, patented high-bulk characteristics. Designed to meet the gloss, opacity, caliper and brightness specifications of heavier papers. For heatset web offset presses. Available in basis weights 32 to 40 lb.

Mainelite® is a No. 5 coated groundwood web offset paper. A No. 5 coated web paper that delivers an ultra-lightweight product with virtually no show-through. It is ideal for projects with postal concerns. For heatset web offset presses. Available in basis weights 26 to 30 lb.

Publication Gloss® is a No. 5 coated groundwood web rotogravure grade. Available in 30 to 50 lb. basis weights. Long known in the industry as a very smooth, high print gloss sheet, Publication Gloss gives you the advantage of lightweight with superb printability.

Rotocote® is a No. 5 coated groundwood web rotogravure paper. This grade is known for consistency, reliability and great printed results. Images are faithfully reproduced with accuracy and vibrancy. For rotogravure web presses. Available in basis weights 26 to 40 lb.

Rotocote® Plus is a No. 5 coated groundwood web rotogravure paper and the best No. 5 coated web paper in the industry. Known for outstanding smoothness, gloss and opacity. It is the perfect paper for magnificent color reproduction, roll after roll. For rotogravure web presses. Available in basis weights 30 to 34 lb.

Rotocote® HB is a No. 5 coated groundwood web rotogravure paper. This grade uses a proprietary high-bulking process to deliver the same quality and printability standards of heavier-weight papers. For rotogravure web presses. Available in basis weights 32 to 34 lb.

Strygloss® is a supercalendered SCA offset paper. This grade delivers reliability and consistency, again and again. For heatset web offset web presses. Available in basis weights 28 to 40 lb.

Reyem® Super is a supercalendered SCA gravure paper. This grade offers a combination of brightness, print quality, runnability and value. The ideal choice when printing volumes are great and budgets are tight. For rotogravure web presses. Available in basis weights 28 to 40 lb.

Insert® Offset is a supercalendered SCB web paper. This grade leads the industry in its category with reliable on-press performance. For heatset web offset presses. Available in basis weights 28 to 40 lb.

Insert® Roto is a supercalendered SCB web paper. This grade yields excellent print reproduction and runnability. Known to be the best in its category. For rotogravure web presses. Available in basis weights 28 to 40 lb.

COMMERCIAL PRINTING PAPERS

International Paper's growth through acquisition has created not just a bigger commercial printing paper business but a better one. It has a renewed focus on simplifying the world of commercial printing. Our strength provides a great depth of capabilities, more flexibility, a strong portfolio of brands and the ability to respond faster to customer needs. We simplified our manufacturing process to provide consistent high quality and streamlined our product offering for easier selection. The result is a focused, more efficient organization, providing more value to our commercial printing customers... the one source to call for The Best Ideas on Paper™.

The following is a list of our portfolio of brands and products.

ACCENT®

The Accent line of opaques delights customers with the combination of quality, availability and service. With minimal dot gain, alkaline longevity and, of course, outstanding opacity, this opaque grade is the perfect accent for promotions, direct mail, financial data, annual reports, manuals and digital printing. Accent Opaque is the clear value in opaques.

A brief overview of the Accent line follows:

Accent® Opaque is available in text basis weights from 40 to 100 lb. in three finishes: smooth, vellum and lustre. It is available in a 92 brightness, blue-white shade and a soothing warm white. Accent Opaque is guaranteed to run on small and large offset presses, copiers, laser and inkjet printers and plain paper fax machines. It is also available with 30 percent recycled content and a 90 brightness on a making order basis only.

Accent® Opaque Cover is available in 65 and 80 lb. cover weights and in three finishes: smooth, vellum and super smooth. It is available in a 92 brightness, blue-white shade and a soothing warm white. Accent Opaque Cover possesses the strength for easy folding, scoring, diecutting and perforating. It is guaranteed for use in electronic imaging equipment that accepts cover weight papers. It is also available with 30 percent recycled content and a 90 brightness on a making order basis only.

Accent® Opaque Digital is available in text basis weights from 45 to 70 lb. and cover weights of 65 and 80 lb. in a smooth, digital finish with a 92 brightness and blue-white shade. It is available in the SKUs that digital printers need most in standard and non-standard sizes both in grain short and long for extra crisp folds. Accent Opaque Digital offers outstanding performance on print-on-demand machines, digital presses, digital office equipment, copiers, laser and inkjet printers, plain paper fax machines and offset presses.

CAROLINA® PAPERS

The Carolina line of coated one-side and coated two-side covers has always been known as a leader in Coated Bristols.

Outstanding brightness, smoothness and consistency have been some of the characteristics that have helped Carolina maintain world class recognition. Available in rolls, sheets and cartons, Carolina offers one of the widest selections of Coated Bristols in the industry.

A brief description of the Carolina offering follows:

Carolina® C2S Cover is designed as an economical alternative to other more expensive No. 2 and No. 3 enamel covers, and provides the excellent printability on both sides that typifies a Carolina product. With its neutral white shade and high brightness, Carolina C2S Cover will make colors pop and images come to life — on both sides of the sheet! Carolina C2S Cover is also suitable for varnish, UV coating, film lamination, extrusion coating, foil stamping and embossing. It can be used for book, magazine and catalog covers, game cards, calendars, direct mail pieces and return postcards.

Carolina® Web Cover is a coated two-side cover specifically designed for heat-set web offset printing. While the special coatings provide a smooth, glossy surface and comparable side-to-side printing results, they also allow optimal press performance on today's high-speed heat-set web offset printing presses. This economical substitute for more expensive web covers will deliver outstanding results in four-color process work, and is suitable for varnish, UV coatings, foil stamping and embossing, and will score, fold and diecut cleanly.

Carolina® C1S Cover is a full density, solid bleached sulfate sheet. The exceptional double coating process of Carolina C1S Cover enhances surface smoothness and printability. Carolina C1S Cover is manufactured to a neutral white shade with high brightness and gloss.

Carolina C1S Cover is printable through both sheet-fed and web processes, and is also suitable for after-print enhancements such as varnish, UV coating, film lamination, extrusion coating, hot foil stamping and embossing.

Carolina C1S Cover is suitable for cover applications that require strength, durability and longevity.

Carolina® C1S Ultra Light Cover is a low density, solid bleached sulfate sheet. The surface is double coated on machine and calendered to provide a superior, printing surface. The lower density characteristics of Carolina C1S Ultra Light Cover makes it an economical alternative for many printing needs. Its neutral white shade and high brightness and gloss allow Carolina C1S Ultra Light Cover to be printed by any process, and is suitable for all after-print enhancements. Outstanding results can be achieved on book covers, greeting cards, brochures, postcards, calendars, magazines covers, catalog covers, tags, folders, magazine inserts and direct mail pieces.

Carolina® C1S Blanks are a medium density, solid bleached sulfate sheet. While significantly higher in caliper and basis weight than cover stocks, Carolina C1S Blanks provide a good, glossy printing surface, but with more stiffness and rigidity. Typical end uses for Carolina C1S Blanks include posters, tags, tickets, point of purchase displays, mobiles, heavyweight covers and specialty advertising pieces.

Carolina® C1S Low Density Blanks are a low density, cost effective alternative to heavier, more expensive blanks. Although being lower in basis weight than medium density products, Carolina C1S Low Density Blanks are specifically designed for superior printability and possess excellent diecutting, scoring, folding and gluing qualities. After-print enhancements, like varnish, UV coating, foil lamination, foil stamping and embossing, highlight the versatility of Carolina C1S Low Density Blanks.

Envirocote™ C1S Cover is a recycled grade similar to Carolina C1S Cover, but contains a minimum of 10 percent post-consumer fiber. Its double-coated top side provides a glossy surface suitable for demanding reproductions. The neutral white shade and high brightness allow book covers, brochures and greeting cards to come to life — all at an economical price.

SPRINGHILL® PAPERS

Springhill Papers was a premier brand of papers in the early 1900s, and it has stood the true test of time. Today, Springhill includes a wide color palette of opaque offsets and uncoated bristols, and it is known for its runnability and economy. A listing of the Springhill family follows.

Springhill® Opaque Offset Colors and **Cover**, available in 13 popular colors, provide consistent quality and opacity along with great runnability. Springhill Opaque Offset Colors and Cover are available in a wide variety of stock sizes along with special make sheets and rolls. These colors match the Springhill Uncoated Bristols Color line, providing economical choices for applications such as the text pages of brochures, price lists, coupons, newsletters, direct mail and other demanding applications where color is preferred. All colors contain 30 percent post-consumer fiber.

Springhill® Vellum Bristol Cover is a dual-purpose cover stock for both bristol and cover applications with 20 percent recycled post-consumer fiber. It has the bulk and feel of a premium cover stock for both bristol and cover applications and without the high price. The sheet's superior formation and dimensional stability offer fast drying capabilities and excellent ink holdout for beautiful full-color reproductions with remarkable depth and color tone. It is excellent for embossing, diecutting, scoring and folding and can be used for brochures, direct mail, greeting and mailing cards, tent cards, menus, announcements and jobs with special finishing and binding requirements. The grade is offered in white and 11 colors.

Springhill® Vellum Bristol Cover *PLUS* is a multi-purpose version of Springhill Vellum Bristol Cover and contains 20 percent recycled post-consumer fiber. It is suitable for all offset duplicators plus high-speed xerographic copiers/laser printers that accept higher basis weight papers. Available in rolls and cutsizes, this is the perfect sheet for tent cards, booklet covers, folders, presentation covers, posters, menus and point-of-purchase materials.

Springhill® Index contains 20 percent recycled post-consumer fiber and is the choice when outstanding printability, strength, stiffness, durability and bulk are the attributes you need. The smoother, hard surface provides superior ink holdout, durability, snap and resilience. The product line also comes in white and eight colors that match Springhill Offset and Vellum Bristol Cover colors. This grade runs effortlessly through scoring, folding, embossing and diecutting equipment. Applications include index systems, file cards, case records, charts, ruled forms, menus, brochure covers, direct mail pieces, counter displays, folders and envelopes.

Springhill® Index *PLUS* is a multi-purpose version of Springhill Index and contains 20 percent post-consumer fiber. It is perfect for all offset duplicating needs and performs well on high-speed copiers/laser printers that accept higher basis weight papers. Excellent for dividers, certificates, charts, inserts, fliers, price lists and mailers.

Springhill® Tag is the strongest of the bristols. It is ideal for printed pieces that must be handled frequently or set up as point-of-purchase displays. The blue-white shade provides excellent image reproduction and it is one of only two tag product offerings on the market that come in colors — seven in all, including white. Tag colors match across other grade lines. Tag offers more stiffness and durability than Index and is the right choice for jobs that will be reused such as time cards, intercompany envelopes, menus, counter and store displays, tags, folders, catalog covers and postcards.

WILLIAMSBURG™

Williamsburg Offset is a trusted offset sheet characterized by its dependable print quality and consistent runnability on press. The attractive blue-white shade provides excellent print contrast. It is available in smooth and vellum finishes as well as 75 lb. hi-bulk Return Postcard. Our cutsize products, Williamsburg Offset *PLUS*, carry the Electronic Imaging Guarantee. Williamsburg Offset is also available with 30 percent post-consumer recycled fiber. It is ideal for letterheads, newsletters, manuals, direct mail, brochures, flyers, catalogs and posters. Williamsburg Offset is a more perfect union between paper and printer.

OFFICE PAPERS

HAMMERMILL® PAPERS

HammerMill produces a broad line of high-performance printing and business papers, everything from premium papers for impressive image-building material to the economy grades you need when budgets are tight.

Always known as an innovator, HammerMill continues to take a leadership role in the development of new products that meet the needs of emerging technologies and the needs of its customers. For example, HammerMill offers opaque and cover papers that run efficiently through the new digital electronic imaging technologies as well as traditional offset presses. It also produces special papers developed specifically for desktop publishing, color copying and other high-tech applications. Multi-purpose sheets made by HammerMill run efficiently through virtually any non-impact printer, copier or small offset press. Colors that range from traditional business hues to sizzling contemporary shades are offered, plus, a wide choice of the recycled grades that are growing in popularity.

MULTI-PURPOSE PAPERS

HammerMill Fore® DP is a multi-purpose sheet with special characteristics that assure top-notch performance on copiers, laser printers and small offset presses. Ideal for offset-printed letterheads and other materials that are first offset-printed and then run through laser and other non-electronic printers. White plus 20 popular colors make it easy to add zest to mailers or color code forms and manual sections. Colors contain 30 percent total recovered fiber, 30 percent post-consumer fiber.

HammerMill Tidal DP™ is a budget-priced, multi-purpose sheet that has the consistent quality needed for no-jam trips through copiers, both large and small, plus the bulk and stiffness required for trouble-free small offset press printing. Delivers dependable performance at a remarkably reasonable cost.

HammerMill Xero/Lasercopy™ has a fine-tuned smoothness that precisely reproduces laser printer, copier and small offset press images, in addition to a bright white and watermark that gives work a VIP look. Ideal for reports and other communications that call for an extra touch of quality.

Other multi-purpose papers include HammerMill Bond, Brite-Hue® by HammerMill, Regalia™ by HammerMill and Via™ by HammerMill.

PREMIUM ELECTRONIC IMAGING PAPERS

HammerMill Color Copy Paper – PhotoWhite was developed specifically for color laser copiers and other electronic color imaging equipment. This ultra bright (96) PhotoWhite sheet has the exceptional formation and smoothness required for unparalleled color reproduction. Its laser-optimized surface picks up every hue and color nuance, producing an exact copy of the original. Available in 28 lb. text and a new PhotoNatural shade. Also available in 60 and 80 lb. Cover.

HammerMill Jet Print was specifically formulated for inkjet printers that use plain paper. Special surface characteristics minimize ink absorption, maximize drying time, prevent feathering and assure sharp, well-defined images. Also performs dependably in copiers, laser printers and plain paper fax equipment. Ideal for presentations, reports, graphs, letters and other premium documents.

HammerMill Jet Print Ultra-Matte is a matte coated sheet with a high-contrast white that delivers high quality prints with crisp black-and-white images and brilliant color. Ideal for monochrome documents with spot color applications. The print side is clearly identified to guarantee ease of printing.

HammerMill Jet Print Ultra-Gloss is a premium gloss coated sheet designed expressly for use on all high-resolution desktop color inkjet printers. Extra heavy weight stands up to heavy ink coverage for superior color prints. Perfect for high quality color presentations and reports.

HammerMill Laser Bond is a premium business paper that offers the prestige of a 25 percent cotton content and watermark. Offers all the advantages of a premium writing paper without the premium price. Ideal for letterheads and other image builders. Laser optimized to assure dependable performance on laser and water-based inkjet printers and offset presses. Available in three dignified business shades.

HammerMill Laser Plus is the graphic arts laser paper. Level, ultra smooth surface faithfully reproduces 300 to 1,200 dpi resolution and delivers sharp, clear, high contrast copies suitable for pre-press proofing and camera-ready masters. Special additive on one side (imprinted for easy identification) prevents bleed-through when adhesive or wax is applied before pasteup.

HammerMill Laser Print is a very level, ultra-smooth surface that maximizes image clarity. No "hills and valleys" to distort dot resolution. Ideal for "premium publishing." Makes presentations, proposals and other high-profile electronically printed documents look their best. Available in 24, 28 and 32 lb. text. Also available in 60 lb. Cover.

TEXT AND COVER PAPERS

Brite-Hue® by HammerMill, Text and **Cover** is an economical text and cover with 17 electrifying cross-matched colors, 10 of which contain 20 percent post-consumer fiber. These high-voltage sheets give 110 percent every time you give them a job to do. Ideal for bulletins, posters, envelope stuffers and other work that demands immediate attention. Seven matching colors in Brite-Hue Laser/Copy.

Via™ **by HammerMill, Text** and **Cover** is a versatile and value-oriented text and cover system that offers flawless print fidelity in 11 neutral colors and eight contemporary colors in five functional finishes at a budget-pleasing price. Tight dimensional stability for impeccable image reproduction with minimal dot gain. Via contains 20 percent post-consumer fiber and is acid-free for added archival quality which extends the life of the document. Via has the strength necessary for easy folding, scoring, diecutting and perforating and is ideal for memos, envelopes, invitations, business cards, annual reports, menus and tent cards.

Regalia™ **by HammerMill, Text** and **Cover** is a premium text and cover paper that denotes elegance and luxury, yet carries an affordable price tag. Outstanding formation and printing surface for superior ink hold out and print quality. The perfect choice for complete business communication packages and sales promotion programs. Three elegant shades of white all containing 20 percent post-consumer fiber are cross-matched to Regalia Writing.

GREAT WHITE®

The demands on today's business papers can be downright menacing. Fortunately, there is Great White® Recycled Multi-Purpose Paper. With 30 percent post-consumer fiber content, Great White reduces the flow of waste paper to landfills and meets the Presidential Executive Order regarding recycled content fiber levels. Smooth-running, like a non-recycled sheet, Great White glides through high-speed copiers/printers, plain paper faxes, laser printers, inkjet printers and offset presses alike. And, Great White's 84 brightness ensures high contrast, presentation quality output of both text and graphics. Plus, Great White is available in our unique Express Pack® with innovative flip-down panel to help simplify handling, reduce the risk of back injury and speed print production.

Great White® Multi-Purpose is a multi-purpose business paper that meets current minimum reclaimed content standards established by the Environmental Protection Agency. Delivers top-notch performance in copiers, laser printers, plain paper fax machines and small offset presses.

INTERNATIONAL PAPER PRINTING PAPERS

International Paper Papers are economical, multi-purpose

sheets designed to run efficiently through copiers, laser printers and plain paper fax machines.

International Paper Relay MP is a precision-sheeted, 84 brightness dual-purpose sheet with outstanding runnability. Economically priced, it is suitable for both moderate and high volume xerographic systems and offset duplicating applications.

International Paper Recycled Relay MP is a recycled dual-purpose sheet suitable for high-speed copiers, inkjet printers, offset duplicators and laser printers. It contains a minimum of 20 percent post-consumer fiber. With an 84 brightness, it is an excellent choice for copiers, manuals and forms.

HAMMERMILL PREMIUMS

Via™ by HammerMill, Writing is a premium watermarked, economically priced writing paper that is available in two weights, 11 colors and three finishes — smooth, linen and laid. Via Writing, combined with Via Text and Cover, provides a platform to meet vital needs for everyday business papers — from letterhead and business cards to brochures and presentation folders. In addition, all recycled papers in this line contain 20 percent post-consumer fiber. Via Writing is acid free for added archival quality that extends the life of the document.

Regalia™ by HammerMill, Writing is a premium writing paper that will help you produce outstanding communications that convey an image of elegance and luxury yet keep costs in line. Regalia Writing, combined with Regalia Text and Cover, is ideal for corporate identity packages, direct mail and sales promotion materials. Three elegant shades of white all containing 20 percent post-consumer fiber are cross-matched to Regalia Text and Cover. Regalia Writing is available in either watermarked or unwatermarked sheets.

HammerMill Bond continues a tradition of excellence that began with the introduction of the original watermarked HammerMill Bond in 1912. Reformulated to meet the demands of today's technologies, HammerMill Bond is a No. 1 sulfite sheet containing 30 percent post-consumer fiber and is a multifunctional paper designed to run on the newest digital electronic imaging equipment as well as traditional printing presses, copiers, and laser and inkjet printers. Moves easily through the entire production procedure, from initial design to final product, delivering flawless images regardless of the imaging process. Available in 92 brightness in Writing and Rippletone finishes.

Brite-Hue® by HammerMill, Laser/Copy is ideal for announcements, newsletters, direct mail — anything that requires immediate attention. Multi-purpose versatility. Zips through copiers, laser printers and small offset presses delivering clear, well-defined images regardless of the production method. Brite-Hue Laser/Copy is available in seven sizzling colors that are cross-matched to Brite-Hue Text and Cover.

Printed by Merrill/Daniels, Everett, Massachusetts
Project coordination by Vicki Tyler and Ashley Futrell, International Paper
Cover design: Michael Coulson and Dick Pepper

Cover: Carolina® Coated Cover, C1S, 10-pt.
Text: Hammermill® Opaque, Smooth, 60 lb.
Pages 79-90: Zanders® Mega Gloss, 80 lb.

Printed in U.S.A